Ecoregions

Springer

New York
Berlin
Heidelberg
Barcelona
Budapest
Hong Kong
London
Milan
Paris
Santa Clara
Singapore
Tokyo

Dedicated to the memory of my son
Matthew G. Bailey
(1968–1998)

Robert G. Bailey

Ecoregions

The Ecosystem Geography of the Oceans and Continents

Illustrations by Lev Ropes

With 106 illustrations, with 55 in color.

Springer

Robert G. Bailey
Ecosystem Management
USDA Forest Service
3825 East Mulberry Street
Fort Collins, CO 80524
USA

Cover: Two ecoregions in juxtaposition: the Rocky Mountains rising above temperate steppes (replaced by wheat fields) of the Great Plains near Longmont, Colorado. Photograph by John Kieffer, reproduced by permission.

Library of Congress Cataloging-in-Publication Data

Bailey, Robert G., 1939–
    Ecoregions : the ecosystem geography of the oceans and continents
    Robert G. Bailey.
      p.   cm.
    Includes bibliographical references (p.   ) and index.
    ISBN 0-387-98305-8 (hardcover : alk. paper), — ISBN 0-387-98311-2
  (pbk. : alk. paper)
    1. Biotic communities—Classification.   2. Biogeography.
  I. Title.
  QH540.7.B345   1998
  577.8′2—dc21   97-26384

Printed on acid-free paper.

Acquisitions Editor: Robert C. Garber
Production coordinated by University Graphics, Inc., and managed by Natalie Johnson; manufacturing supervised by Jacqui Ashri.
Typeset by Matrix Publishing Services.
Printed and bound by Walsworth Publishing Co., Marceline, MO.
Printed in the United States of America.

9 8 7 6 5 4 3 2 1

ISBN 0-387-98305-8 Springer-Verlag New York Berlin Heidelberg (Hardcover) SPIN 10635027
ISBN 0-387-98311-2 Springer-Verlag New York Berlin Heidelberg (Softcover) SPIN 10634990

Preface

Most environmental concerns cross boundaries. Borders that separate countries, ecosystems, or jurisdiction of regulatory agencies are not respected by problems such as air pollution, declining anadromous fisheries, forest diseases, or threats to biodiversity. To address these problems, environmental planners and decision makers must consider how geographically related systems are linked to form larger systems. Issues that may appear to be local will often require solutions at the landscape and regional scale—working with the larger pattern, understanding how it works, and designing in harmony with it.

Following this reasoning, my task was to develop a geographical, ecologically based system that would classify the natural ecoregions of the Earth and plot their distribution. The project built on work published by John Crowley in 1967 (Crowley 1967). In 1989, I published a terrestrial ecoregions map at a scale of 1:30,000,000 (Bailey 1989). A simplified, reduced-scale version appears in my book *Ecosystem Geography* (Bailey 1996) and the 19th edition of *Goode's World Atlas*. The influence of such ecoregion mapping efforts on research and planning efforts has been considerable. For example, the National Science Foundation's Long-Term Ecological Research program and The Nature Conservancy's ecoregional planning programs are taking place within the framework of this map. The Sierra Club announced in 1994 (*Sierra* March–April 1994) a "critical ecoregions" program designed to protect and restore twenty-one regional ecosystems in the United States and Canada. In another effort, the U.S. Forest Service has adopted ecoregion units as part of an ecologically based mapping system to support ecosystem management and assessment.

Understanding the continental systems requires a grasp of the ocean systems that exert enormous influence on terrestrial climatic patterns.

This book is unique in the extended treatment (see Chapters 2 and 3, and Plate 1) of oceanic ecoregions.

Ecoregions is intended to provide detailed descriptions, illustrations, and examples that will assist the user of the ecoregion maps in interpreting them. It amplifies the necessarily brief descriptions of the ecoregion units that appear in the legends to the maps. However, description without reference to genesis or origin soon becomes dull and tiresome—terms that unfortunately characterize much of the ecological literature on regions. A major objective of this book, therefore, is to suggest explanations of the mechanisms that act to produce the world pattern of ecoregion distribution, and to consider some of the implications for land use. The global extent of this book and its maps dictates that its ecoregion classification scheme be kept simple as possible, recognizing only principal ecoregion types. Where regional studies require additional detail, numerous additional subdivisions as needed can be added within an ecoregion type. For an example of detailed regional studies of ecoregions that follow this principle of creating subdivisions within the recognized world types, see *Description of the Ecoregions of the United States* (Bailey 1995).

Ecoregions was written for several audiences. In addition to the environmental planners and decision makers mentioned at the beginning of this Preface, I hope that the increasing visibility of scale- and system-based science in ecological and environmental research will bring this work to the attention of workers in those fields. In addition, students and instructors should find the ecoregion approach useful in courses ranging from environmental planning to biogeography and ecosystem or landscape ecology.

I am indebted to Preston E. James, Günter Dietrich, John J. Hidore, Arthur N. Strahler and Alan H. Strahler, Heinrich Walter, and J. Schultz for their own wonderful and insightful books.

I would like to acknowledge John M. Crowley, who began the work of global ecosystem regionalization. Recognition should also go to Chris Risbrudt, Director of Ecosystem Management Coordination in the Washington Office of the U.S. Forest Service, for his support. My thanks to Lev and Linda Ropes for helping me explain and illustrate the ideas in this book. The maps were made by Jon Havens. I am also indebted to my wife, Susan Strawn Bailey, who made some of the drawings and provided encouragement and advice. As always, it has been a pleasure to work with Rob Garber at Springer-Verlag.

Fort Collins, Colorado   Robert G. Bailey

Contents

Frontispiece. The Colorado Plateau, Utah–Arizona, illustrates how a temperate steppe ecoregion is subdivided by the arrangement of surface features (which modify the regional climate) into sand dune, rock, and shrubland sites. Photograph by John S. Shelton, reproduced with permission.

Introduction

Environmental problems are best addressed in the context of geographic areas defined by natural features rather than by political or administrative boundaries. For example, the state of Colorado in the western United States is neatly and abruptly divided into two areas with dramatically different ecological, climatological, and land-use characteristics: the eastern plains and western mountains (Figure 1.1). Furthermore, both areas extend beyond Colorado's borders.

More and more we recognize that the natural resources of an area do not exist in isolation. Instead, they interact so that use of one affects another (Figure 1.2). This understanding has led entities responsible for the management of public—and increasingly, private—land to delineate and manage **ecosystems**,[1] rather than individual species or individual resources such as timber or range.

This approach recognizes that the Earth operates as a set of interrelated systems within which all components are linked. A change in one component causes a change within another with corresponding geographic distributions, as when certain vegetation and soil types occur together with certain types of climate. The regions occupied by tropical rainforest, for example, are found in tropical wet climates, and the underlying soils in the rainforest tend to be latosols (Oxisols)[2] (Figure 1.3).

Ecosystems occur on many geographic scales. We can recognize ecosystems of different size, from oceans to frog ponds, and vast deserts to pockets of soil. The smaller systems are embedded, or nested, within larger systems. The larger systems are the environments of those within, controlling their behavior. By understanding the large forces that create macroscale ecosystems, we can predict how management practices will affect smaller component systems. For example, on a macroscale, the continents are embedded in the ocean systems that

Figure 1.1. Abrupt rise of the Colorado Front Range above the smooth surface of the North American Great Plains, looking north near Colorado Springs. Photograph by T.S. Lovering, U.S. Geological Survey.

control them through their influence on climatic patterns. A hierarchical ordering of the scales of ecosystems, from macro to micro, is presented in detail in my related book, *Ecosystem Geography*.

Concepts of Ecosystem Regions, or Ecoregions

Any large portion of the Earth's surface over which the ecosystems have characteristics in common is called an ecosystem region, or **ecoregion**. Plates 1 and 2 show that areas with similar ecosystems are found in similar latitudinal and continental locations. Therefore, the distrib-

Figure 1.2. Formation of gullies due to overgrazing, erosion, and increased runoff near Canberra, Australia. Grazing has eliminated the grass cover, reducing the retention of rainwater and facilitating the concentration of runoff. Photograph by Stanley A. Schumm, U.S. Geological Survey.

ution of ecoregions is not haphazard; they occur in predictable locations in different parts of the world and can be explained in terms of the processes producing them. For instance, temperate continental ecoregions in the Northern Hemisphere are always located in the interior of continents and on the leeward, or eastern, sides; thus the northeastern United States is in some ways similar to northern China, Korea, and northern Japan (Figure 1.4 and Plate 2).

Because of this predictability, we can make assumptions about ecological features such as vegetation type that can be transferred across similar ecoregions of the same continent, or analogous ecoregions on different continents. Because data can be extended reliably to analo-

Tropical wet climates

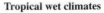

Tropical rainforests

Latosolic soils

Figure 1.3. Spatial correspondence in the tropics among broad categories of climate, vegetation, and soils. Climate from Trewartha; vegetation after Eyre, Küchler, and others; soils based on numerous sources, including Soil Conservation Service. From *Physical Elements of Geography*, 5th ed., by Glenn T. Trewartha, Arthur H. Robinson, and Edwin H. Hammond, Frontispiece, Plate 5, Plate 6. Copyright © 1967 by McGraw-Hill Inc. Reproduced with permission of The McGraw-Hill Companies.

gous sites within an ecoregion, we may greatly reduce data sampling and monitoring.

Need for a Comparative System of Generic Regions

Some schemes of classifying ecosystems have been based on the intuitive recognition of homogeneous-appearing regions, without considering the controlling forces that differentiate them. Using such methods each ecoregion is considered unique, unrelated to other regions. These are nothing more than "place name regions," such as the Great Plains of North America or the high altiplano of Bolivia, instead of being based on criteria that define what type of region each is.

Figure 1.4.
Global pattern
of the warm
continental
ecoregions.

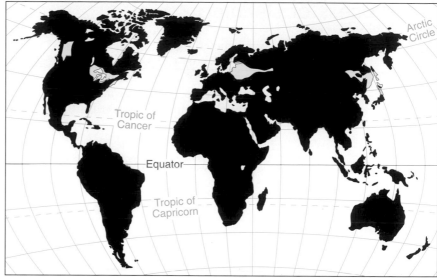

Warm continental zones

As a result of this, analogous regions in different continents or oceans may not be defined in the same way. Such inconsistency makes it difficult to exchange environmental information. Regions defined without specifying the factors upon which they were based are difficult for others to scrutinize or confirm. The results are therefore difficult to communicate convincingly. In this book, I use a more explicit approach in which regions are studied on the basis of comparable likenesses and differences. Such explicit methods require us to consider the physical factors that underlie ecosystem differentiation.

Understanding the processes involved in ecosystem (ecoregion) differentiation provides a basis for selecting significant criteria: those that are responsible for creating the range of ecoregion types found on the Earth. The purpose of this book is to describe and explain the character and arrangement over the Earth of the major ecosystem types, and the causes behind those patterns.

The face of the Earth could yield a nearly infinite variety of regional ecosystem types, each defined by application of different criteria. But not all homogeneous areas are of equal significance. We are seeking meaningful types by the identification of correspondence between spatial patterns. That is, the manner in which the variations within one ecosystem component correspond to the variations within another component, particularly in ways that affect process. For example, all steep slopes have shallow soils and are susceptible to erosion.

The Process of Defining

The fundamental question facing all ecological mappers is: How are the boundaries of systems to be determined? The first, large-scale divisions of the Earth's surface are, quite obviously, the land masses and the water areas, where ecological processes take place in quite a different manner. But how do we determine the distribution of ecosystems within each?

First we analyze, on a scale-related basis, the controlling factors that differentiate ecosystems. We use significant changes in those controls as boundary criteria.

The Role of Climate

Energy is the prime driving force and controller of conditions on Earth. The three main sources of energy are (1) solar radiation, providing heat and light, (2) the kinetic energy of the rotation and orbit of the Earth, and (3) internal forces of both heat and kinetic energy. Figure 1.5 shows the first two sources. Internal forces will be discussed in Chapter 4.

As a result of the way the Earth revolves about the sun and rotates on its axis, the low latitudes or tropics receive more solar radiation than do middle and higher latitudes.[3] Only about 40% as much solar energy is received above the poles as above the equator. To balance

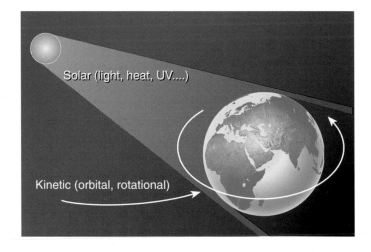

Figure 1.5.
Fundamental sources of energy that control conditions on Earth. Base from Mountain High Maps®. Copyright © 1995 by Digital Wisdom, Inc.

this energy, there is a large-scale transfer of heat poleward, which is accomplished through atmospheric and oceanic circulation. The frictional effects of the rotating Earth's surface on air flow cause these circulations to be relatively complex. Nevertheless, the solar energy and atmospheric and oceanic circulations are distributed over the Earth in an organized fashion. These controls, in turn, produce recognizable world patterns of temperature and precipitation, the two most important climatic elements.

Climate, as a source of energy and water, acts as the primary control for ecosystem distribution. As climate changes, so do ecosystems, as a petrified forest lying in a desert attests (Figure 1.6).

Figure 1.6. Petrified forest that now lies in a desert zone in Arizona. Photograph by D.B. Sterrett, U.S. Geological Survey.

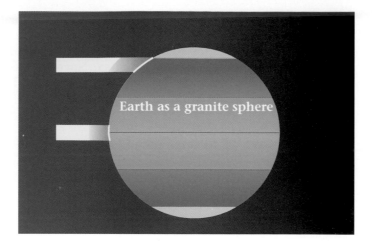

Figure 1.7. Latitudinal climatic zones that would result if the Earth were simply a granite sphere with an atmosphere.

Figure 1.8. Pattern of land, water, and ice near the South Pole. Base from Mountain High Maps®. Copyright © 1995 by Digital Wisdom, Inc.

Macroclimate

If the Earth had a homogeneous surface (either land or water), there would be circumferential zones of climate resulting from variation in the amount of solar radiation that reaches different latitudes (Figure 1.7). These large climatic zones, or **macroclimates**, would be arranged simply in latitudinal bands, or east–west belts. They would owe their differentiation to the varied effects of the sun, instead of the character of the surface.

However, the Earth's surface is rather heterogeneous (Figure 1.8), divided first into large masses of land (the continents), water (the oceans), and ice (the polar regions). Each modifies the otherwise simple macroclimates. We first consider the oceans in the next chapter.

CHAPTER 2

Oceanic Types and Their Controls

Oceans occupy some 70% of the Earth's surface and extend from the North Pole to the shores of Antarctica. There are great differences in the character of the oceans, and these differences are of fundamental importance, both in the geography of the oceans themselves, and to the climatic patterns of the whole Earth. The surface of the ocean is differentiated into regions, or zones, with different hydrologic properties resulting from unevenness in the action of solar radiation and other phenomena of the climate of the atmosphere.

Ocean hydrology and atmospheric climate are both dynamic and closely interrelated. Ocean hydrology may be defined as the seasonal variation in temperature and salinity of the water. It controls the distribution of oceanic life. We use ocean hydrology to differentiate regional-scale ecosystem units and to indicate the extent of each unit. With continental ecosystems, atmosphere is primarily responsible for ecoregions, but in the ocean, the physical properties of the water determine ecoregions, not the atmosphere.

Factors Controlling Ocean Hydrology

To establish the hydrographic zones of the world's oceans, we must determine the boundaries of the zones. Our approach to this task is to analyze those factors that control the distribution of zones, and to use significant changes in those controls as the boundary criteria. This distribution in the character of the oceans is related to the following controlling factors:

Latitude

Heating depends predominantly on latitude. If the Earth were covered completely with water, thereby eliminating the deflection of currents by land masses, there would be circumferential zones of equal ocean temperature. The actual distribution forms a more complicated pattern (Figure 2.1). Water temperatures range from well over 27°C in the equatorial region to below 0°C at the poles. Areas of higher temperature promote higher evaporation and therefore higher salinity.

At low latitudes throughout the year and in middle latitudes in the summer, a warm surface layer develops, which may be as much as 500 m thick. Below the warm layer, separated by a thermocline, is a layer of cold water extending to the ocean floor. In arctic and antarctic regions, the warm surface water layer disappears and is replaced by cold water lying at the surface. The boundary between water types, called the **oceanic polar front**, is nearly always associated with a convergence of surface currents, which are discussed below.

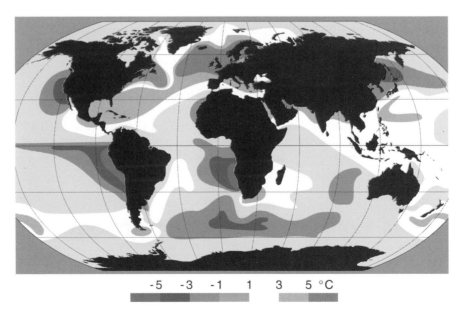

-5   -3   -1   1   3   5 °C

Figure 2.1. Ocean surface temperatures expressed as a deviation from what they would be on a hypothetical globe covered only by water. From *General Oceanography: An Introduction* by G. Dietrich, Chart 3. Copyright © 1963 by John Wiley & Sons, Inc. Reprinted by permission of John Wiley & Sons, Inc. and Gebrüder Borntraeger Verlag.

Major Wind Systems

Prevailing surface winds set surface currents in motion and modify the simple latitudinal arrangement of thermal zones. The most important of these winds, centered over the ocean basins at about 30° in both hemispheres, are know as **oceanic whirls** (Figure 2.2). They circulate around the **subtropical high-pressure cells**, created by subsiding air masses (see Figure 4.4, p. 36). Lying between the subtropical belts and the equator are easterly winds, known as **trade winds**. Near the equator the trade winds of both hemispheres converge to form a low pressure trough, often referred to as the **intertropical convergence zone**, (ITC). On the poleward side of the subtropical highs is a belt of westerly winds, or **westerlies**. Generally, wind-driven ocean currents move warm water toward the poles and cold water toward the tropics, in a clockwise direction in the Northern Hemisphere and a counterclockwise direction in the Southern Hemisphere (Figure 2.3).

The effects of these movements may be seen in Figure 2.1. The average sea temperature on the coast of southern Japan, washed by the

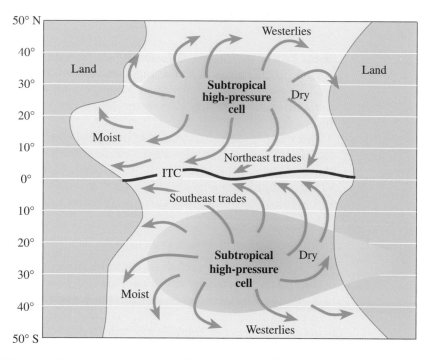

Figure 2.2. Over the oceans, surface winds spiral outward from the subtropical high-pressure cells, feeding the trades and the westerlies. These drive circular flow on the surface of the oceans. Adapted from Strahler (1965), p. 64; reproduced with permission.

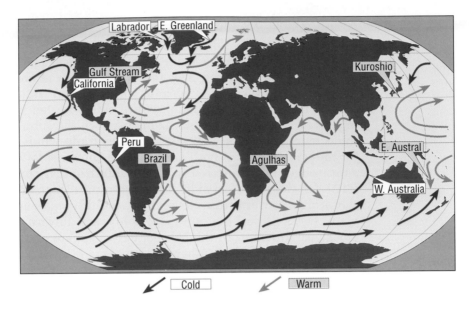

Figure 2.3. The major oceanic surface currents are warm when flowing poleward and cold when flowing from the poles.

warm Kuroshio Current, is nearly 8° warmer than that in southern California, in the same latitude, but bathed by a cool current reinforced with **upwelling**. These are areas where ocean currents tend to swing away from the continental margins and cold water wells up from underneath.

The Earth's rotation, tidal swelling, and density differences further contribute to the dynamics of the oceans, both at the surface and well below. Distribution and diversity of ocean life are subsequently affected. Abundance is generally higher in areas of cooler temperatures and areas of upwelling of nutrients (Figure 2.4).

Precipitation and Evaporation

The amount of dissolved salts, called **salinity**, varies throughout the oceans. It is affected by the local rates of precipitation and evaporation. Heavy rainfall lowers the surface salinity by dilution; evaporation raises it by removing water. Average surface salinity worldwide is about 35 parts per thousand (Figure 2.5).

The highest salinities in the open sea are found in the dry, hot tropics, where evaporation is great. Nearer the equator salinities decrease because rainfall is heavier. In the cooler middle latitudes, salinities are relatively low because of decreased evaporation and increased precipitation. Surface salinities are generally low in the Arctic and Antarc-

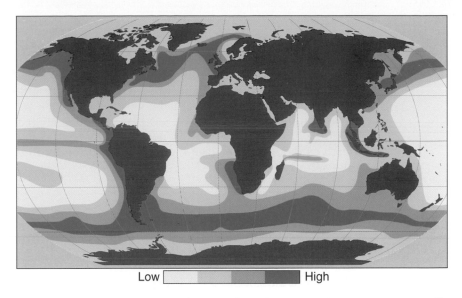

Low [ ] High

Figure 2.4. The distribution of zones of productivity in the oceans is generally related to areas of cool water and upwelling. From Lieth (1964–1965).

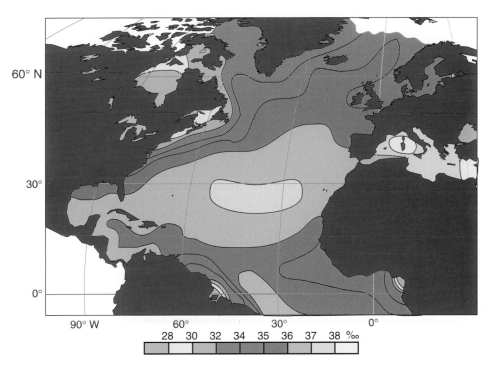

Figure 2.5. Surface salinities in the North Atlantic and adjacent waters are affected by the amounts of precipitation and evaporation. Modified from Dietrich (1963); in *Physical Elements of Geography* by Glenn T. Trewartha, Arthur H. Robinson, and Edwin H. Hammond, p. 392. Copyright © 1967 by McGraw-Hill Inc. Reproduced by permission of The McGraw-Hill Companies.

tic waters because of the effect of melting ice, which dilutes the sea water. In coastal waters and nearly enclosed seas, the salinity departs greatly from this pattern. In hot, dry seas of the Mediterranean and Middle East, the salinity is greater because the water is subject to strong evaporation and cannot mix readily with the open ocean. Near the mouths of large rivers, and in nearly enclosed seas fed by large rivers, such as the Baltic Sea (see Figure 2.5), dilution by fresh water reduces the salinity.

Types of Hydrologic Regions, or Oceanic Ecoregions

The interaction of the oceanic macroclimates and large-scale ocean currents determine the major hydrologic regions with differing physical and biological characteristics. These regions are defined as oceanic ecoregions. Their boundaries follow the subdivision of the oceans into hydrographic regions by Günter Dietrich (1963), although some modification has been made to identify regions with high salinity and the zonal arrangement of the regions. Dietrich's oceanographic classification and their ecoregion equivalents are summarized in Tables 2.1 and

Table 2.1. Natural regions of the oceans[a]

Dietrich group and types	Ecoregion equivalents
B **Boreal regions**	**Polar domain** (500)
Inner boreal (Bi)	Inner polar division (510)
Outer boreal (Bä)	Outer Polar division (520)
W **Westerly drift ocean regions**	**Temperate domain** (600)
Poleward of oceanic polar front (Wp)	Poleward westerlies division (610)
Equatorward of polar front (Wä)	Equatorward westerlies division (620)
R **Horse latitude ocean regions (R)**	Subtropical division (630)
	High salinity subtropical division (640)[b]
F **Jet current regions (F)**	Jet stream division (650)
M **Monsoon ocean regions**	
Poleward monsoons (Mp)	Poleward monsoon division (660)
	Tropical domain (700)
Tropical monsoon (Mt)	Tropical monsoon division (710)
	High salinity tropical monsoon division (720)[b]
P **Trade wind regions**	
With poleward current (Pp)	Poleward trades division (730)
With westerly current (Pw)	Trade winds division (740)
With equatorward component (Pä)	Equatorward trades division (750)
A **Equatorial ocean regions (Ä)**	Equatorial countercurrent division (760)

[a]From the Dietrich system (1963).
[b]Not recognized by Dietrich; from Elliott (1954).

Table 2.2. Definitions and boundaries of the Dietrich system

P	(*Passat* in German) persistent westerly setting currents
Pä	with 30-degree equatorward component
Pw	predominantly westerly set
Pp	with 30-degree poleward current
Ä	(*Äquator* in German) regions of currents directed at times or all year to the east
M	(*Monsun* in German) regions of regular current reversal in spring and autumn
Mt	low-latitude monsoon areas of little temperature variation
Mp	mid to high (poleward) latitude equivalents of large temperature variations
R	(*Ross* in German) at times or all year marked by weak or variable currents
F	(*Freistrahlregionen* in German) all-year, geostropically controlled narrow current belts of mid-latitude westerly margins of oceans
W	(*Westwind* in German) marked by somewhat variable but dominantly east-setting currents all year
Wä	equatorward of oceanic polar front (convergence)
Wp	poleward of oceanic polar front (convergence)
B	at times or throughout the year ice covered, in Arctic and Antarctic seas
Bi	entire year covered with ice
Bä	winter and spring covered with ice

2.2. Dietrich's system was used with similar results by Hayden et al. (1984) in their proposed regionalization of marine environments.

Dietrich's Oceanographic Classification

Dietrich's classification takes into account the circulation of the oceans, the temperature and salinity, and indirectly, the presence of major zones of upwelling. The motion of the surface is emphasized because of its influence on temperature. Salinity is generally higher in areas of higher temperature, and therefore, higher evaporation. Marine organisms are usually more abundant in cold waters, which makes the water appear more green.

There are seven main groups of hydrologic zones. Six are based on ocean current, while the polar group is not. Four of these are subdivided into types based on current direction, latitude, and duration of pack ice. All together there are twelve regional divisions.

Dietrich (1963) describes them briefly. In the equatorial region is a belt with currents directed toward the east. These are the currents of the equatorial region, or the *Ä* group. On the low-latitude margins of the equatorial region are the currents of the trade wind regions, or the *P* group, where persistent currents move in a westerly direction. The *P* group is subdivided into three types, with a poleward current (*Pp*), a westerly current (*Pw*), and an equatorward component (*Pä*). Poleward from the trade winds is the horse wind region, or *R* group, where weak currents of variable direction exist. Still farther poleward is the region of west wind drift, or *W* group, where variable easterly currents prevail. Here, two subdivisions are recognized: poleward of the oceanic front (*Wp*) and equatorward of the polar front (*Wä*). In the very high latitudes are the polar regions, or the *B* group, which are covered by ice. The duration of ice determines the subdivision: outer polar (*Bä*) in winter and spring, which is covered with pack ice, and inner polar (*Bi*), which is covered with ice the entire year. The monsoon regions, group *M*, are subdivided into poleward monsoon (*Mp*) and tropical monsoon (*Mt*). A final subdivision is the jet stream region, or *F* group, where strong, narrow currents exist as a result of discharge from trade wind regions.

Distribution of the Oceanic Regions

I have combined and rearranged the twelve regional divisions to identify oceanic regions. The resulting ecoregion **divisions** are shown diagrammatically in Figure 2.6 as they might appear in a hypothetical ocean basin. On the diagram, twelve kinds of regional divisions are recognized. They range from the inner polar division at high latitudes to the equatorial countercurrent division at low latitudes. Two of them are further subdivided (not shown on diagram) on the basis of high salinity following work by Elliott (1954).

I further simplified this classification of oceanic regions, using some of Schott's ideas (1936, as reported by Joerg 1935), by grouping the divisions in larger regions, called **domains** (see Figure 2.6). This recognizes the fact that the oceanic regions are arranged in latitudinal belts that reflect the major climatic zones. In both hemispheres there are three contrasting types of water, differing in temperature, salinity, life forms, and color. They are separated along major lines of convergence where surface currents meet. In the high latitudes are the *polar waters*, characterized generally as low in temperature, low in salinity, rich in plankton, and greenish in color. These currents also frequently carry drift ice and icebergs. In the low latitudes are the *tropical waters*, generally high in temperature, high in salinity, low in organic forms, and blue in color. In between lie the so-called *mixed waters* of the temperate, middle latitudes.

As is the case in all phenomena on the Earth's surface that are directly or indirectly related to climate, there is a tendency toward regularity in the pattern of arrangement of the hydrologic zones. However, the distribution of these various kinds of ocean water and the convergences that separate them is not strictly latitudinal. The east coasts of the continents are bathed by poleward-moving currents of

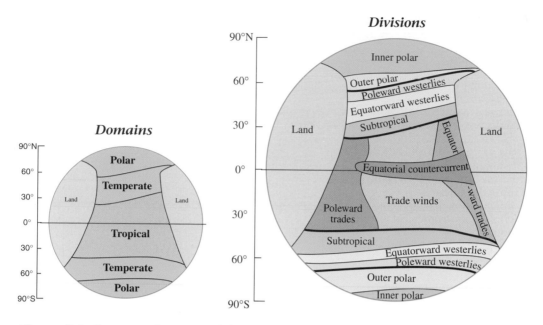

Figure 2.6. Ecoregion domains and divisions in a hypothetical ocean basin. Compare with ocean map, Plate 1.

tropical water. The west coasts at the same latitudes are bathed by colder waters moving from the poles. Because of the configuration of the ocean basins, the warm, east-coast currents of the Atlantic and the Pacific Oceans are best developed in the Northern Hemisphere, and the cold, west-coast currents in the Southern Hemisphere. As mentioned above, where currents tend to swing offshore, cold water wells up from below.

In my system, there are four domains and fourteen divisions, plus the continental shelf area where the shallow waters (<200 m) are interpreted as shallow variations of the ecoregion involved. Plate 1 (inside back cover) shows the distribution of the oceanic ecoregion domains and divisions. Table 2.3 presents a rough estimate of the area contained in each region and its percentage of the total ocean area.

Each domain/division occurs in several different parts of the world that are broadly similar with respect to physical and biological characteristics. For example, the equatorial trades division (750) occurs only on the west sides of large continents and is set off from the adjacent low-latitude oceanic waters by lower temperatures, somewhat wider variations of temperatures, and lower salinity.

We will now treat the characteristics, extent, and subdivisions of each of the regions distinguished in the next chapter. Their character-

Table 2.3. Approximate area and proportionate extent of oceanic ecoregions

	km²	Percent
500 Polar domain	50,601,000	14.13
510 Inner polar division	20,305,000	5.67
520 Outer polar division	30,296,000	8.46
600 Temperate domain	136,620,000	38.15
610 Poleward westerlies division	17,977,000	5.02
620 Equatorward westerlies division	56,868,000	15.88
630 Subtropical division	52,929,000	14.78
640 High salinity subtropical division	1,217,000	0.34
650 Jet stream division	4,583,000	1.28
660 Poleward monsoon division	3,043,000	0.85
700 Tropical domain	118,714,000	33.15
710 Tropical monsoon division	9,239,000	2.58
720 High salinity tropical monsoon division	2,685,000	0.75
730 Poleward trades division	18,299,000	5.11
740 Trade winds division	54,325,000	15.17
750 Equatorward trades division	14,790,000	4.13
760 Equatorial countercurrent division	19,373,000	5.41
Shelf	52,177,000	14.57

ization provides further information on the principles that have been applied in drawing the boundaries in Plate 1. The descriptions of the characteristics are drawn and summarized from several sources; the most important is Dietrich (1963), with supplementary information by Elliott (1954) and James (1936).

CHAPTER 3

The Ecoregions of the Oceans

500 Polar Domain

At times, or during the winter, ice of the Arctic or Antarctic Oceans covers the regions of the polar domain. They are characterized, in general, by ocean water that is greenish, low in temperature, low in salt content, and rich in small, sometime microscopic plant and animal organisms, know as **plankton**. The duration of ice provides a basis for division into (a) an *inner polar zone* (division) covered by ice for the entire year, and (b) an *outer polar zone* (division) where, with a 50% probability, ice is encountered during winter and spring. Figure 3.1 shows the global locations of the polar ecoregions of the oceans.

510 Inner Polar Division

In the arctic region, the inner polar region includes the deep arctic basin and the northern passages of the Canadian Archipelago; in the antarctic region, it includes a narrow belt around the antarctic ice shelf. This is the region that Dietrich (1963) designates as *Bi*. Although the region is always covered with ice, the ice itself is in continuous motion, clockwise around the poles. In the Northern Hemisphere this area is free of icebergs, because no glaciers touch this region. Hummocked pack ice forms impenetrable ridges with intermittent leads and openings here and there, due to severe ice pressure. In contrast, the central parts of the inner polar region are relatively flat, pack-ice fields.

The Arctic Ocean, which is surrounded by land masses, is normally covered by pack ice throughout the year, although open leads are numerous in the summer. The relatively warm North Atlantic drift maintains an ice-free zone off the northern coast of Norway. The situation

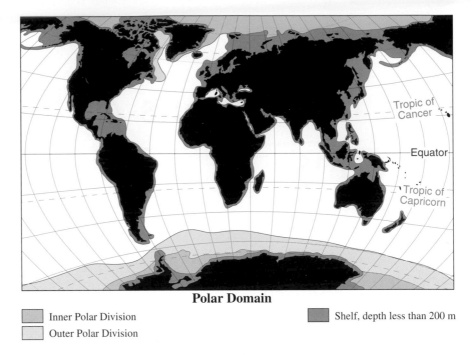

Polar Domain

Inner Polar Division   ▮ Shelf, depth less than 200 m
Outer Polar Division

Figure 3.1. Divisions of the oceanic polar domain.

is quite different in the Antarctic, where a vast, open ocean bounds
the sea ice zone on the equatorward margin. Because the ice flows can
drift freely north into warmer waters, the antarctic ice pack does not
spread far beyond 60° S latitude in the cold season.

Air temperatures are generally below freezing, and the annual variation of air temperature is large. Because of the low air temperatures
and the resultant low capacity of the air to hold moisture, the annual
amount of precipitation is very small, but exceeds evaporation.

520 Outer Polar Division

Adjacent to the inner polar region lies the outer polar region, which
is regularly covered by pack ice during winter in both hemispheres. It
is classified as *Bä* in the Dietrich system. Pack ice does not originate
within this region. It represents ice carried into this region by oceanic
currents from the inner polar region, and forms fields of drifting ice
that will gradually melt. Oceanic currents considerably influence the
position of the boundaries of this region. With the cold East Greenland
Current and the Labrador Current, the polar pack ice is carried far to
the south end of the Grand Banks (46° N). In contrast, the extension
of the warm Gulf Stream system keeps the eastern side of the Norwe-

gian Sea and the southern Barents Sea ice free during the entire year. This allowed convoys from America during World War II to reach the Russian (formerly Soviet) port of Murmansk.

In summer, the hydrologic conditions of the *Bä* regions do not differ considerably from those of the *Wp* regions. This is especially true for the Southern Hemisphere, where variable, predominantly easterly currents are common to both regions. The boundaries between the two regions, as well as the boundary between the two polar regions, favor the development of plankton. In the feeding chain, plankton forms the basis of krill shrimp on which the blue whales and finback whales feed. These relationships explain why in summer the two boundaries become the main feeding grounds of the whales and, consequently, the main whaling grounds.

600 Temperate Domain

This domain comprises the middle latitudes between the poleward limits of the tropics and the equatorward limits of pack ice in winter. Currents in this region correspond to wind movements around the subtropical, high-pressure cells of the atmosphere. These are the so-called "mixed waters" of the middle latitudes. The variable direction of the current determines the divisions: (a) a *poleward westerlies zone* (division) with cold water and sea ice that is poleward of the oceanic polar front; (b) an *equatorward westerlies zone* (division) that has cool water and is equatorward of the oceanic polar front; (c) a *subtropical zone* (division) with weak currents of variable directions; (d) a *high salinity, subtropical zone* (division) characterized by excess evaporation over precipitation; (e) a *jet-stream zone* (division) characterized by strong, narrow currents that exist during the entire year as the result of discharge from trade wind regions; and (f) a *poleward monsoon zone* (division) of high latitudes with reversal of current (connected with large annual variations in surface temperature). A global map of the temperate ecoregions of the oceans is shown in Figure 3.2.

610 Poleward Westerlies Division and 620 Equatorward Westerlies Division

These regions are characterized by variable, predominantly easterly currents during the entire year. The polar boundary is the zone that, during winter, is covered permanently or frequently with ice of the polar seas. In winter, east-traveling cyclones are strongly developed and produce storm zones. These include the "roaring forties" between 40°

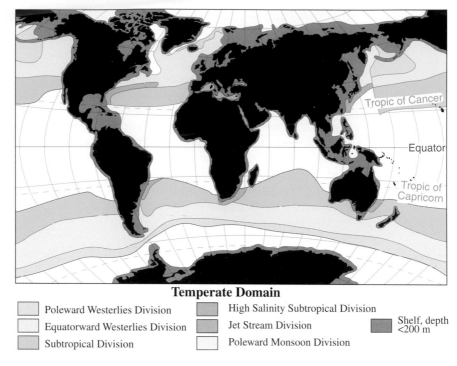

Temperate Domain

☐ Poleward Westerlies Division	▨ High Salinity Subtropical Division	
☐ Equatorward Westerlies Division	▨ Jet Stream Division	▨ Shelf, depth <200 m
▨ Subtropical Division	☐ Poleward Monsoon Division	

Figure 3.2. Divisions of the oceanic temperate domain.

and 50° in the Southern Hemisphere, as well as storm zones in the North Pacific and North Atlantic Oceans.

In the *W* regions in the Dietrich system, precipitation falls during all seasons and higher and more frequently in fall and winter. Precipitation exceeds evaporation, decreasing the surface salinity. The oceanic polar front lies within the *W* regions. In the Southern Hemisphere, it coincides with the zones of strongest westerly winds. At the sea surface it separates the cold-water sphere from warm water. Along this front, the spreading of ice of the polar seas generally reaches its equatorial limit. This provides a basis to subdivide the *W* regions into a *Wp* region (poleward of the polar front) and a *Wä* region (equatorward of the polar front).

Deep-reaching mixing processes are present in these regions. The associated high-nutrient concentrations provide an opportunity for plankton to develop abundantly. Plankton provides nutrition for great numbers of commercial fish, especially on the continental shelves. The shelves provide other sources of nutrition and also serve as spawning grounds. For this reason, the Grand Banks, the banks west of Greenland, around Iceland, and the Faroes, as well as the shelf of Europe have become the main grounds of high-sea fishing. Oceanic islands in these regions, such as the Crozet Islands in the South Indian Ocean, have antarctic birds and mammals which exist because of food chains that start in these seas (Figure 3.3).

630 Subtropical Division

Between the region of the trade wind currents, P in the Dietrich system, and the region of the westerlies, lies a region of transition designated by the symbol R. In this region, weak currents of variable direction exist. This region is associated with the subtropical high, which is associated with weak winds—called the *horse latitudes* at about 25° to 30° north and south of the equator. The horse latitudes are said to be so called from the throwing overboard of horses in transport from Europe to North America if a ship's passage was delayed by calms.

Figure 3.3. Sea elephants and penguins on the Crozet Islands, South Indian Ocean. Photograph by Douglas Mawson; from the American Geographical Society Collection, University of Wisconsin–Milwaukee Library.

Trade winds and westerlies flow around the interior region of the horse latitudes. This causes an accumulation of light surface water in the center of the eddy. A deep, homogeneous, warm, top layer is established, which is also very saline because of the excess of evaporation over precipitation (see Figure 2.5, p. 15). No other region in the world ocean has higher temperature and salinity.

The R regions are also characterized by an extremely low nutrient content because plankton have consumed the nutrients. Therefore, the biomass assumes its smallest value in the surface waters, such as in the Sargasso Sea in the central R region of the North Atlantic Ocean. In contrast, the absolute maximum of organic production in the ocean has been found in the region of upwelling off southwest Africa (see Figure 2.4, p. 15).

As a consequence of very low plankton content, these regions are distinguished by extremely clear, transparent water, which shows a deep cobalt blue color. They are generally smooth seas because storms rarely touch them.

640 High Salinity Subtropical Division

The high salinity regions are not designated as a separate variety by Dietrich, which Elliott (1954) designates as type H. They have been incorporated into the system presented in this book. These regions are represented by the Mediterranean Sea, the Red Sea, and the Black Sea. Continental influences are extremely strong because the individual units are almost landlocked and water exchange with the oceans is very slow. Each unit shows strong individuality, but nevertheless certain characteristics permit grouping them into one major type. These common characteristics are: almost landlocked position, extremely strong continental influence, wide range and variation of temperatures, wide range of salinities, wide variation of air temperature, great range of precipitation, and generally counterclockwise surface circulation. All are characterized by excess of evaporation over precipitation; making them highly saline (see Figure 2.5, p. 15).

650 Jet Stream Division

Along the west sides of the oceans in low latitudes, the equatorial current turns poleward, forming a warm current paralleling the coast, indicated by the symbol F in the Dietrich system. These discharge currents are known as the Gulf Stream and Kuroshio in the Northern Hemisphere, and as the Brazil Current, East Austral Current, and Agulhas Current in the Southern Hemisphere. They bring higher than average temperatures along these coasts. The high velocity of these currents tends to cause a cross circulation. At the left side of the Gulf Stream, for example, the cross circulation brings water rich in nutrients from deeper layers to subsurface layers. This is associated with an abundance of plankton, causing the greenish-blue color of this adjacent water.

660 Poleward Monsoon Division

Dietrich classified the poleward monsoon waters as *Mp*. These regions occur mainly in east Asiatic waters, where regular changes in the direction of the monsoon winds in spring and fall cause a reversal of the surface currents. The season of the winter monsoon lasts from November to March. During this time, under the influence of cooling, a strong atmospheric high develops over Asia. The air in the lowest layers flows out over the east Asiatic marginal seas (i.e., the Sea of Japan, the Sea of Okhotsk, and the Bering Sea) from north to northwest. During the summer, from May to September, the flows are reversed on shore toward the continent.

The offshore winds of the winter monsoon carry cold continental air over the seas, whereas the onshore winds of the summer bring warm, oceanic air masses into these regions. Due to these atmospheric influences, the sea surface temperatures have the greatest annual variations found in any ocean. In North Korean and Manchurian waters, variations of over 20°C, sometimes even 25°C, are found.

In winter, the water temperature drops to the freezing point. Even Vladivostok, Russia, on the latitude of Florence, Italy, does not remain ice free. In spring and summer, when the maritime air masses saturated with water vapor arrive monsoon-like over these cold oceanic areas, persistent sea fogs develop.

Comparative low salinities are caused by heavy runoff from the land, great excesses of precipitation over evaporation, and inflow of low salinity, polar continental waters.

700 Tropical Domain

This is characterized by ocean water that is generally blue, high in temperature, high in salt content, and low in organic forms. It is divided into (a) *tropical monsoon*, with regular reversal of the current system (connected with small annual variations in surface temperature); (b) *high salinity, tropical monsoon*, with alternating currents; (c) *poleward trades*, with a strong velocity directed toward the poles; (d) *trade winds*, with current moving toward the west; (e) *equatorward trades*, with a strong velocity directed toward the equator, and where currents tend to swing offshore and cold water well up from below; and (f) *equatorial countercurrents*, where currents are directed at times or during the entire year toward the east. The global extent of the tropical ecoregions of the oceans is presented in Figure 3.4.

710 Tropical Monsoon Division

Located over the North Indian Ocean and the waters around southeast Asia, is a region affected by monsoonal winds, which Dietrich designates as *Mt*. In winter, the winds blow from the north and northeast. In this season, the hydrologic conditions resemble those of a region of trade wind currents. The surface water flows toward the west and becomes more saline along its way because the dry continental air does not produce precipitation and evaporation is high.

Annual temperature variation is quite small; however, currents, winds, salinities, and precipitation show marked seasonal change because of the monsoon winds, which cause a reversal of the currents with the season.

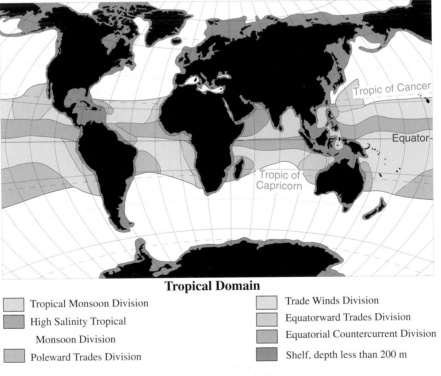

Tropical Domain

Tropical Monsoon Division	Trade Winds Division
High Salinity Tropical	Equatorward Trades Division
Monsoon Division	Equatorial Countercurrent Division
Poleward Trades Division	Shelf, depth less than 200 m

Figure 3.4. Divisions of the oceanic tropical domain.

720 High Salinity Tropical Monsoon Division

This type is restricted to the Arabian Sea, where dry winds bring about very low precipitation, strong evaporation, and resulting high salinities. In type 710, designated as type *E* by Elliott, salinities are considerably lower because of the monsoon rains and high runoff.

730 Poleward Trades Division

North and south of the equator are the currents of the trade wind belts, covering roughly the zones lying between 5° and 30° N and S. Its poleward boundaries follow approximately the mean stand of the longitudinal axes of the subtropical high-pressure cells (see Figure 4.5, p. 37). Air moving equatorward from the cells is deflected by the Earth's rotation to turn westward, with currents running the same direction the entire year. Turning of the current (see Figure 2.2, p. 13) is controlled partly by the turning of the trade winds and partly by the distribution of the continents. Those winds deviate the currents from the zonal direction in the eastern and western regions.

In the west of the division are regions of the trade wind currents with components directed toward the poles. These regions, which Dietrich designated as *Pp*, are distinguished by anomalously high surface temperatures, which, together with the high evaporation, contribute to instability of the atmosphere and precipitation. If orographic rain also falls on the coast, the amounts are sufficient to develop lush vegetation on oceanic islands, such as the Polynesian Islands and the continental coast of Brazil, south of Bahia, as well as northeastern Australia.

740 Trade Winds Division

In these regions, the North and South Equatorial Currents, which Dietrich classified as *Pw*, move westward uniformly and persistently. The uniformity of current corresponds to uniformity of wind and weather. These regions have little precipitation and, since evaporation is high, the sea surface is highly saline. Because the region is free of divergences and the annual variation of surface temperature is small, there is no vertical mixing to renew nutrients from deeper layers. Because the plankton population is, therefore, low, only a few higher marine organisms exist here.

High sea-surface temperatures, which are over 27°C in these latitudes, are important in the development of tropical storms, which originate in this area. Warming of air at low levels creates instability and predisposes the area toward the formation of storms. Once formed, the storms move westward through the trade wind belt. They are by far the most violent storms on Earth and are know as *typhoons* in East Asia, as *Mauritius hurricanes* in the South Indian Ocean, and as *hurricanes* in the West Indies.

750 Equatorward Trades Division

This type consists of currents known as the Canaries and Benguela Currents in the North and South Atlantic Oceans, as California and Peru (Humbolt) Currents in the North and South Pacific Oceans, and as the West Australia Current in the South Indian Ocean. In the Dietrich system, areas of this type are designated as *Pä*. Since they flow from higher to lower latitudes, they carry water that is colder than average for the corresponding latitudes. These temperature differences become larger because the winds that blow parallel to the coast deflect surface water seaward, thereby causing **upwelling**. This brings cold water, abundant in nutrients, to surface layers. It contributes to the extraordinary development of plankton. The abundance of plankton shows in the green coloring of the sea water, in contrast to the cobalt blue of the neighboring regions of the horse latitudes. The great abundance of plankton produces large amounts of fish in these areas, attracting endless numbers of sea birds. They are responsible for the formation of guano deposits at these coasts. When this upwelling is disturbed or replaced by a motion directed toward the pole (the so-called El Nino at the north Peruvian coast, which is not associated with upwelling but carries warm, nutrient-depleted, equatorial water), then masses of fish, and subsequently birds, die.

In general, surface temperatures are lower than air temperatures, especially in the vicinity of coasts. Hence, formation of lasting and frequent fogs, known under the name Garua at the Peruvian coast, is common. Low rainfall is the result of the stabilization of the air mass over the cold water surface. These areas have the lowest precipitation on the globe, including inland deserts. Oceanic islands in these regions, such as the Cape Verde Islands in the North Atlantic Ocean and the Galapagos Islands in the South Pacific Ocean, have desert-type climates. The coastal deserts of Namib in southwest Africa and of Atacama in north Chile are located at approximately the same latitude adjacent to these cold currents.

Another consequence of the low surface temperature of these regions is the fact that no coral reefs are found in these areas. Their development requires not only clear water but also temperatures of at least 20°C in the coldest month of the year (Figure 3.5).

760 Equatorial Countercurrent Division

The equatorial currents are separated by an equatorial countercurrent and are designated *Ä* by Dietrich. This condition is well developed in

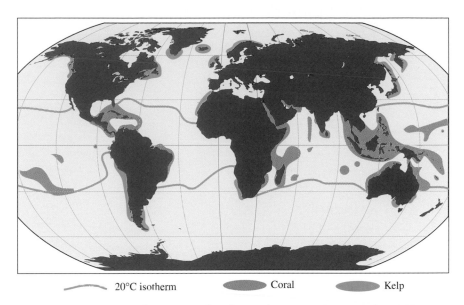

Figure 3.5. Coral reefs are most abundant in the oceans between the 20°C mean annual surface water temperature isotherm, whereas kelp is most abundant outside that zone. After K.H. Mann, Seaweeds: their productivity and strategy for growth, *Science* 182, p. 976, 1973 and J.W. Wells, Coral reefs, *Geological Society of America Memoir 67* 1, p. 630, 1957; in *Ecology of World Vegetation* by O.W. Archibold, figure 12.17, p. 405. Copyright © 1995 by Chapman & Hall. Reproduced with permission.

the three oceans. This region, lying roughly 5° S and 5° N latitude, coincides with a trough of low pressure, the intertropical convergence zone (see also Chapter 4, p. 35, and Chapter 8, p. 105). The northeast and southeast trade winds come together toward this trough. It is a zone of variable winds and calms, or the **doldrums**. The zone contains large amounts of moisture, and cloudiness and frequent precipitation are common. The heavy precipitation lowers the salinity of the sea surface considerably.

Upwelling at the poleward side of the countercurrent results in ascending water masses rich in nutrients. As it reaches the surface, an abundant plankton population develops (see Figure 2.4, p. 15). Like the areas of upwelling in the eastern region of the trade winds currents, the equatorial countercurrent zones are characterized by greenish color and an abundance of fish.

CHAPTER 4

Continental Types and Their Controls

The ecoregions on the Earth's land masses are arranged in predictable patterns and are causally related to macroclimate, i.e., the climate that lies just above the local modifying irregularities of landform and vegetation. These macroclimates are regularly arranged with reference to several controlling factors.

The Controls of Macroclimate

The controls of macroclimatic may be grouped under three headings: latitude, continental position, and altitude.

Latitude

As mentioned previously, if the Earth had a homogeneous surface, circumferential zones of macroclimate would result from the variation in solar radiation and the resultant atmospheric circulation. These zones would be divided along lines of latitude (see Figure 1.7, p. 8).

Thermally Defined Zones.   The actual distribution of land and sea forms a more complicated thermal distribution (Figure 4.1). The thermal limits for plant growth determine boundaries of these zones. For example, trees in Eurasia and America cannot grow beyond about 70° latitude.

We can delineate three major thermally defined zones: (1) a winterless climate of low latitude, (2) a temperate climate of mid latitudes with both a summer and winter, and (3) a summerless climate of high latitude. In the winterless climate, no month of the year has a mean monthly

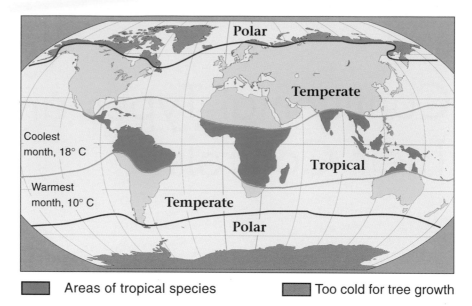

Areas of tropical species   Too cold for tree growth

Figure 4.1. Zones determined by thermal limits for plant growth. From Strahler (1965), p. 103; redrawn with permission.

temperature lower than 18°C. The 18°C isotherm approximates the position of the boundary of the poleward limit of plants characteristic of the humid tropics, such as palms. In the summerless climate, no month has a mean monthly temperature higher than 10°C. The 10°C isotherm closely coincides with the northernmost limit of tree growth, separating the regions of boreal forest (tayga[4]) from the treeless tundra.

If we also consider the annual and diurnal energy cycles, we can differentiate these thermal zones. The relative amplitudes of annual and diurnal energy cycles vary in each region (Figure 4.2). Within the tropics, the diurnal range is greater than the annual range. Within temperate zones, the annual range exceeds the diurnal range, although the diurnal can be very large. Within the polar zones, the annual range is far greater than the diurnal range.

Moisture-Defined Zones.   Life-giving precipitation is generally higher in areas of higher temperature and, therefore, higher evaporation. It ranges from about 200 cm in the equatorial region to about 30 cm at the poles (Figure 4.3).

Precipitation and runoff also follow a zonal pattern, generally decreasing with latitude. Near the equator the trade winds converge toward the intertropical convergence zone (ITC). The trade winds moving toward the equator pick up moisture over the oceans and, when lifted in the ITC, yield abundant precipitation.

On top of this, differential heating at different latitudes causes transfer of heat from lower to higher latitudes, partly through circulation of the atmosphere. The result is a series of belts of ascending and subsiding air masses (Figure 4.4). Subsiding air masses of the subtropical high-pressure cells, roughly centered on the tropics of Cancer and Capricorn, have adequate heat but a shortage of moisture. These zones are too dry for tree growth.

Continental Position

The interaction of land and sea modifies the situation. This division of the Earth's surface between land and sea, each with quite different

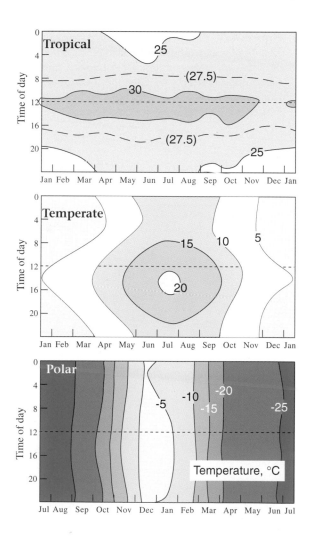

Figure 4.2. Variation in air temperature through the day and through the year in tropical, temperate, and polar zones. Stations are Singapore, Oxford, England, and Mc-Murdo Sound, Antarctica, respectively. From Troll (1966), p. 15–16.

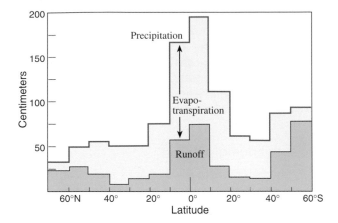

Figure 4.3. Distribution of annual precipitation and runoff amounts averaged by latitudinal zones. The vertical difference between the two lines represents the loss through evapotranspiration. After L'vovich and Drozdov; from *Physical Elements of Geography*, 5th ed., by Glenn T. Trewartha, Arthur H. Robinson, and Edwin H. Hammond, p. 413. Copyright © 1967 by McGraw-Hill, Inc. Reproduced with permission of The McGraw-Hill Companies.

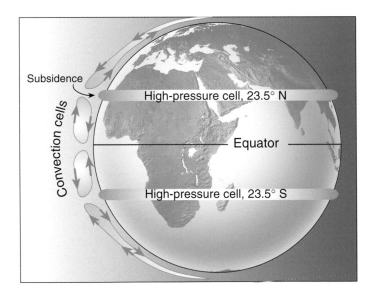

Figure 4.4. The dry zones are controlled by the subtropical high-pressure cells that are caused by subsidence between the atmospheric convection cells. Base from Mountain High Maps®. Copyright © by Digital Wisdom, Inc.

thermal characteristics, results in distinctive differences between marine-influenced and inner-continental climates. For a hypothetical continent of uniform elevation, temperature distribution would look like that shown in Figure 4.5.

In addition, the distribution of land and sea forms complicated but predictable, precipitation patterns—less precipitation over margins of continents bathed by cold water. The dry zone, controlled by the subtropical high-pressure cells, is shifted to the west side of the continents, adjacent to these cold currents (Figure 4.6). These dry zones are too dry for trees and consist of deserts and grasslands. This creates a distribution of dry climatic zones that strongly affect ecosystem distribution.

Combined Latitude (Thermal) and Moisture Considerations.   By combining the thermally defined zones with the moisture zones, we can delineate four ecoclimatic zones or ecoregions: *humid tropical*, *humid temperate*, *polar*, and *dry* (Figure 4.7). They are arranged in a regular, repeated pattern with reference to latitude and position on the continents.

Within each of these zones, one or several climatic gradients may affect the potential distribution of the dominant vegetation. Within the humid tropical zone, for example, we can distinguish rainforests that have year-round precipitation from savannas that receive seasonal precipitation. Thus, we can subdivide the humid tropical zone, based on

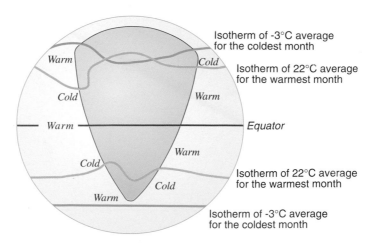

Figure 4.5. Summer and winter isotherms as they might appear on a hypothetical continent, and the temperatures of adjacent waters. From *A Geography of* 2d ed., by Preston E. James, p. 181. Copyright © 1959 by Ginn and Com Reprinted by permission of John Wiley & Sons, Inc.

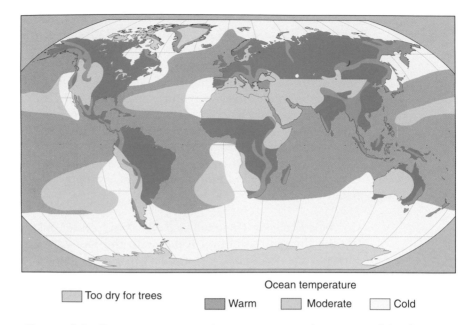

Too dry for trees

Ocean temperature

Warm   **Moderate**   **Cold**

Figure 4.6. Ocean temperatures determine, in part, the position of the dry zones on the continents. After Gerhard Schott; from *A Geography of Man*, 2d ed., by Preston E. James, p. 632. Copyright © 1959 by Ginn and Company. Reprinted by permission of John Wiley & Sons, Inc.

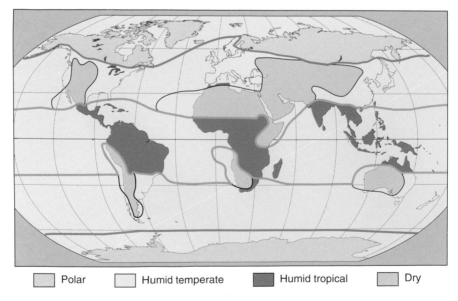

Polar   **Humid temperate**   **Humid tropical**   **Dry**

Figure 4.7. Four ecoclimatic zones of the Earth based on temperature and moisture.

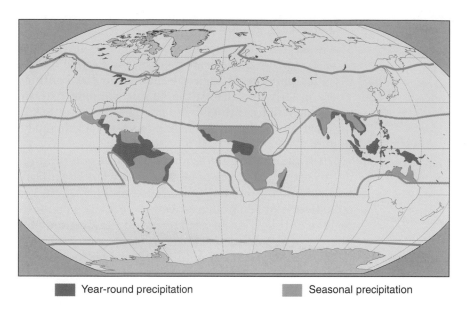

| ■ Year-round precipitation | ■ Seasonal precipitation |

Figure 4.8. Subdivisions of the humid tropical zone based on seasonal moisture distribution.

moisture distribution (Figure 4.8) into *climatic subzones.* We can subdivide the other zones similarly.

Locating the boundaries of broad-scale ecosystems requires taking into account visible and tangible expressions of climate such as vegetation. Generally, each climate is associated with a single **plant formation class** (such as savanna, see Figure 8.5, p. 109; Table 4.1), and is characterized by a broad uniformity both in appearance and in composition of the dominant plant species. Usually a significant correspondence with soils occurs because climate also strongly dominates soil-forming processes.

Table 4.1. Broad plant formations and groups of climates

Formation	Köppen climate group
Tropical rain forest	A (tropical rainy climates)
Tropical desert	B (dry climates)
Temperate deciduous forest	C (warm temperate climates)
Boreal forest	D (snowy-forest climates)
Tundra	E (polar climates)

Modified by Altitude

The arrangement of the ecological zones is largely dependent on latitude. To further complicate matters, the Earth's internal energy causes irregular patterns of high mountains on the continents (Figure 4.9). These modify and distort the simple climatic pattern that would develop on a flat continent. We can see the short-term and small-scale implications of this difference in local meteorological boundary effects (Figure 4.10).

These mountains are arranged without conforming at all to the orderly latitudinal zones of climate. They cut irregularly across latitudinally oriented climatic zones. For example, we find mountains in the cold deserts of Antarctica as well as near the equator (Figure 4.11). The regions of this type do not appear on the diagram showing the generalized global pattern of ecoregions (see Figure 4.14, p. 46) because these features, along with the outlines of the continents, are unique for each land mass.

Mountains have a typical sequence of altitudinal belts, with different ecosystems at successive levels (Figure 4.12). These differ according to the zone in which the mountain is embedded. In other words, altitude produces a predictable variation of the lowland climate, especially in **climatic regime** (i.e., seasonality of temperature and precipitation). The coast ranges of California, for example, experience the same strong seasonal energy variations and a seasonal moisture regime

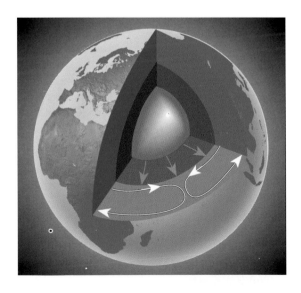

Figure 4.9. The Earth's internal energy sources drive mantle convection and plate tectonics, causing mountain building. Base from Mountain High Maps®. Copyright © by Digital Wisdom, Inc.

Figure 4.10. Cloud formations differentiating a mountain boundary. View west toward the Sangre de Christo Mountains, Colorado. Photograph by Lev Ropes.

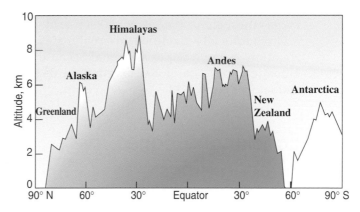

Figure 4.11. The altitude of major mountain areas is generally independent of latitude.

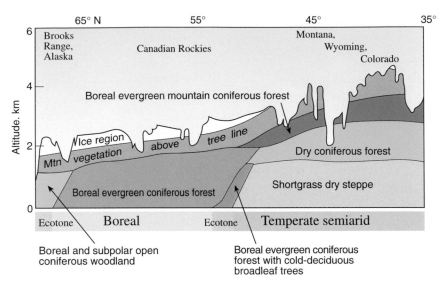

Figure 4.12. Vertical zonation in different ecoclimatic zones along the eastern slopes of the Rocky Mountains in North America. From Schmithüsen (1976), p. 70.

consisting of a dry summer and a rainy winter typical of their neighboring lowlands.

When a mountain occurs in two climatic zones, it produces different vertical zonation patterns. This is shown in Figure 4.12, which compares locations in the Rocky Mountains. In the semi-arid steppe climatic portion, the lowermost zone is a sagebrush basal plain; this is followed by a montane zone of Douglas fir and spruce and fir. Above is the subalpine zone, followed by alpine tundra, and then perennial ice and snow. This sequence of altitudinal zones repeats on mountain ranges throughout the lowland, semi-arid climatic zone.

Orographically modified macroclimates, which exhibit altitudinal zonation, are referred to as **azonal** because they can occur in any ecoclimatic zone. The global distribution of the regions of the type are shown in Figure 4.13.

Types of Ecoclimatic Zones, or Continental Ecoregions

The effects of latitude, continental position, and altitude combine to form the world's ecoclimatic zones, herein referred to as *continental ecoregions*. Their boundaries follow the subdivision of the Earth into climatic zones established by Köppen as modified by Trewartha (Table 4.2)[5], although some modifications have been made to maximize correspondence of the regions with the vegetation (plant formation classes).

The Köppen–Trewartha Classification of Climates

The Köppen–Trewartha classification identified six main groups of climate, and all but one—the dry group—are thermally defined. They are as follows:

Based on temperature criteria

A. Tropical: Frost limit in continental locations; in marine areas 18°C for the coolest month
C. Subtropical: 8 months 10°C or above
D. Temperate: 4 months 10°C or above
E. Boreal: 1 (warmest) month 10°C or above
F. Polar: All months below 10°C

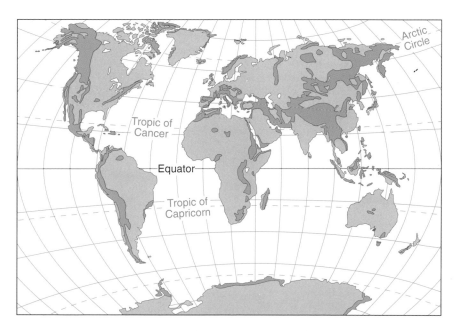

Figure 4.13. The arrangement of orographically modified macroclimates, or highland climates, is distributed as a result of mountain building rather than latitude and therefore cuts across other latitudinally based climates.

Table 4.2. Regional climates[a]

Köppen group and types	Ecoregion equivalents
A **Tropical and humid climates**	**Humid tropical domain** (400)
Tropical wet (Ar)	Rainforest division (420)
Tropical wet-dry (Aw)	Savanna division (410)
B **Dry climates**	**Dry domain** (300)
Tropical/subtropical semi-arid (BSh)	Tropical/subtropical steppe division (310)
Tropical/subtropical arid (BWh)	Tropical/subtropical desert division (320)
Temperate semi-arid (BSk)	Temperate steppe division (330)
Temperate arid (BWk)	Temperate desert division (340)
C **Subtropical climates**	**Humid temperate domain** (200)
Subtropical dry summer (Cs)	Mediterranean division (260)
Humid subtropical (Cf)	Subtropical division (230)
	Prairie division (250)[b]
D **Temperate climates**	
Temperate oceanic (Do)	Marine division (240)
Temperate continental, warm summer	Hot continental division (220)
(Dca)	Prairie division (250)[b]
Temperate continental, cool summer	Warm continental division (210)
(Dcb)	
	Prairie division (250)[b]
E **Boreal climates**	**Polar domain** (100)
Subarctic (E)	Subarctic division (130)
F **Polar climates**	Tundra division (120)
Tundra (Ft)	
Icecap (Fi)	

Definitions and Boundaries of the Köppen–Trewartha System

Ar   All months above 18°C and no dry season.
Aw Same as Ar, but with 2 months dry in winter.
BSh Potential evaporation exceeds precipitation, and all months above 0°C.
BWh One-half the precipitation of BSh, and all months above 0°C.
BSk Same as BSh, but with at least 1 month below 0°C.
BWk Same as BWh, but with at least 1 month below 0°C.
Cs   Eight months 10°C, coldest month below 18°C, and summer dry.
Cf   Same as Cs, but no dry season.
Do Four to seven months above 10°C, coldest month above 0°C.
Dca Four to seven months above 10°C, coldest month below 0°C, and warmest month above
      22°C.
Dcb Same as Dca, but warmest month below 22°C.
E   Up to three months above 10°C.
Ft   All months below 10°C.
Fi   All months below 0°C.

A/C boundary = Equatorial limits of frost; in marine locations, the isotherm of 18°C for
              coolest month.
C/D boundary = Eight months 10°C.
D/E boundary = Four months 10°C.
E/F boundary = 10°C for warmest month.
B/A, B/C, B/D, B/E boundary = Potential evaporation equals precipitation.

[a]Based on the Köppen system of classification (1931), as modified by G.T. Trewartha (1968) and Trewartha et al. (1967).
[b]Köppen did not recognize the prairie as a distinct climatic type. The ecoregion classification system represents it at the arid sides of the Cf, Dca, and Dcb types.

Based on precipitation criteria

B. Dry: Outer limits, where potential evaporation equals precipitation

These groups are subdivided into fifteen types based on seasonality of precipitation or on degree of dryness or cold. They range from the icecaps at high latitudes to the tropical wet climate at low latitudes. Trewartha (1968) describes them briefly. The low latitudes contain a winterless, frostless belt with adequate rainfall. This is the tropical humid climate, or the *A* group. It is subdivided into two types, tropical wet (*Ar*) and tropical wet-and-dry (*Aw*). On the low-latitude margins of the middle latitudes, where winters are mild and killing frosts only occasional, is the subtropical belt, or *C* group. Here two subdivisions are recognized: subtropical dry-summer (*Cs*) and subtropical humid (*Cf*). Poleward from the subtropics is the temperate belt, or *D* group. It contains two types, temperate continental (*Dc*) and temperate oceanic (*Do*). Two subtypes of temperate continental are recognized: a more moderate one with hot summers and cold winters (*Dca*), and a more severe subtype (*Dcb*), located poleward, which has warm summers and rigorous winters. Still farther poleward is the boreal of subarctic belt, the *E* group. It has not been subdivided. In the very high latitudes are the summerless, polar climates (*F* group), subdivided into the tundra climate (*Ft*) and icecap climate (*Fi*). The dry climates, group *B*, are subdivided into semi-arid or steppe type (*BS*) and an arid or desert type (*BW*). A further subdivision separates the hot tropical-subtropical deserts and steppes (*BWh*, *BSh*) from the cold temperate-boreal deserts and steppes (*BWk*, *BSk*) of middle latitudes.

Highland climates, which are low-temperature variants of climates at low elevations in similar latitudes, are designated by the letter *H*.

Distribution of the Continental Regions

Although we can define ecoregions climatically, they are most effectively treated by combining and rearranging the fifteen climatic types to maximize correspondence with major plant formations. Through this process I have mapped the Earth into zones called *ecoregion provinces*, each of which has characteristic ecosystems. This map is based on a world map of natural climate-vegetation landscape types in the *Fiziko-Geograficheskii Atlas Mira* (Gerasimov 1964). Eighty-six such subdivisions are recognized (Bailey 1989), but for simplification I have grouped them into fifteen *divisions*. They range from the tundra at high latitudes to the rainforest at low latitudes. Mountains exhibiting alti-

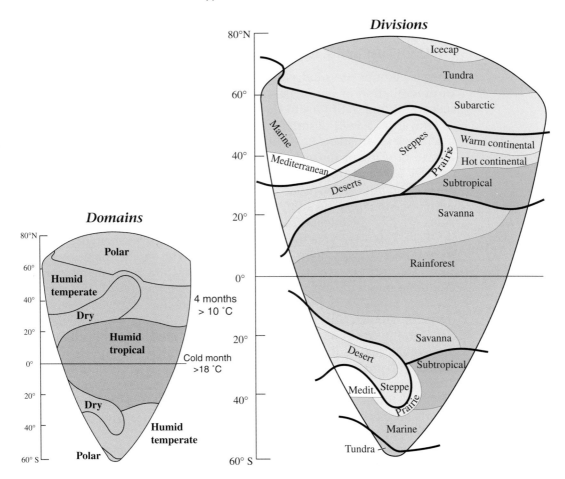

Figure 4.14. Patterns of ecoregions that might occur on a hypothetical continent of low, uniform elevation. Compare with map of continents, Plate 2.

tudinal zonation and having the climatic regime of the adjacent lowlands are distinguished according to the character of the zonation.

We can further simplify this classification of ecosystems by grouping the divisions into four large regions called *domains*. Four such subdivisions are recognized—three are humid zones, thermally differentiated: *polar*, with no warm season; *humid temperate*, rainy with mild to severe winters; *humid tropical*, rainy with no winters. The fourth subdivision, *dry*, is defined on the basis of moisture alone and transects the otherwise humid domains.

These domains can be described briefly. In the very high latitudes lies the polar domain, differentiated on the basis of ice formation and plant development into icecap, tundra, and subarctic tayga divisions. In the mid latitudes is the humid temperate domain of mid-latitude forests, differentiated according to the importance of winter frost into warm continental, hot continental, subtropical, marine, prairie, and mediterranean divisions. In low and middle latitudes is the dry domain, differentiated on the basis of rainfall (steppe versus desert) and winter temperature (cold versus warm) into tropical/subtropical steppe, tropical/subtropical desert, temperate steppe, and temperate desert. In the low latitudes lie the humid tropical domain, differentiated on the basis of rainfall seasonality into savanna and rainforest divisions.

We have identified two principles in the global pattern of ecosystem regions. One is the principle of regularity which relates to those features of the Earth's surface that are associated with climate; the other is the principle of irregularity which relates to all those features associated with land surface forms. The principle of regularity allows us to forecast the kinds of associated features that can be expected at any given latitude and longitudinal position. However, irregularities distort these regular patterns on each specific continent.

We can plot the regular pattern of these regions over the Earth's continents—a pattern that is generalized in Figure 4.14. Because the arrangement of these regions is regular, it is possible to predict the kind of region that will be found in any particular part of the Earth's land areas. Mountains provide the distortions observed on the actual map of the ecoregions in Plate 2 (inside back cover). Mountains exhibiting altitudinal zonation and having the climatic regime of the adjacent lowlands do not appear on our hypothetical continent because they are unique for each land mass.

Each group includes regions in different parts of the world that are broadly similar with respect to surface features and the cover of vegetation. For example, the tropical/subtropical steppe division (310) of the dry domain is found on all continents. These steppes typically are grasslands of short grasses and other herbs with local shrub and woodland. Pinyon–juniper woodland, for example, grows on the Colorado Plateau of the United States (Figure 4.15). These areas may support

Figure 4.15. Sheep passing through pinyon–juniper grazing land of central Utah. Photograph by Robert G. Bailey.

limited grazing, but they are not moist enough for cultivated crops without irrigation.

Table 4.3 lists climate, vegetation, and soil types associated with each zone (division). In Chapters 5 through 8, we describe the characteristics of each of the ecoregion domains and divisions of the continents. These characteristics result from the interplay of surface features, climate, vegetation, soil, water, and fauna. The proportion of the Earth's land area included in each group is shown in Table 4.4.

The descriptions are based on information compiled and summarized from many sources, the most important of which are James (1959), Hidore (1974), Strahler and Strahler (1989), Walter (1984), and Schultz (1995). Climate descriptions are based largely on the Köppen–Trewartha classification, but we also explain them in terms of air masses and frontal zones which are shown in Appendix A. In presenting the

Table 4.3. General environmental conditions for ecoregion divisions

Name of division	Equivalent Köppen–Trewartha climates	Zonal vegetation	Zonal soil type[a]
110 Icecap			
120 Tundra	Ft	Ice and stony deserts: tundras	Tundra humus soils with soilifluction (Entisols, Inceptisols, and associated Histosols)
130 Subarctic	E	Forest-tundras and open woodlands; tayga	Podzolic (Spodosols and Histosols)
210 Warm Continental	Dcb	Mixed deciduous-coniferous forests	Gray-brown podzolic (Alfisols)
220 Hot Continental	Dca	Broadleaved forests	Gray-brown podzolic (Alfisols)
230 Subtropical	Cf	Broadleaved-coniferous evergreen forests; coniferous-broadleaved semi-evergreen forests	Red-yellow podzolic (Ultisols)
240 Marine	Do	Mixed forests	Brown forest and gray-brown podzolic (Alfisols)
250 Prairie	Cf, Dca, Dcb	Forest-steppes and prairies; savannas	Prairie soils, chernozems, chestnut-brown soils (Mollisols)
260 Mediterranean	Cs	Dry steppe; hardleaved evergreen forests, open woodlands and shrub	Soils typical of semi-arid climates associated with grasslands
310 Tropical/subtropical steppe	BSh	Open woodland and semideserts; steppes	Brown soils, and sierozems (Mollisols, Aridisols)
320 Tropical/subtropical desert	BWh	Semideserts; deserts	Sierozems and desert soils (Aridisols)
330 Temperate steppe	BSk	Steppes; dry steppes	(same as BSh)
340 Temperate desert	BWk	Semideserts and deserts	(same as BWh)
410 Savanna	Aw, Am	Open woodlands, shrubs and savannas; semi-evergreen forest	Latosols (Oxisols)
420 Rainforest	Ar	Evergreen tropical rain-forest (selva)	Latosols (Oxisols)

[a]Great soil group. Names in parenthesis are Soil Taxonomy soil orders (USDA Soil Survey Staff 1975). Described in the Glossary, p. 143.

Table 4.4. Approximate area and proportionate extent of ecoregions

	km^2	Percent
100 Polar domain	38,038,000	26.00
110 Icecap division	12,823,000	8.77
M110 Icecap regime mountains	1,346,000	0.92
120 Tundra division	4,123,000	2.82
M120 Tundra regime mountains	1,675,000	1.14
130 Subarctic division	12,259,000	8.38
M130 Subarctic regime mountains	5,812,000	3.97
200 Humid temperate domain	22,455,000	15.35
210 Warm continental division	2,187,000	1.49
M210 Warm continental regime mountains	1,135,000	0.78
220 Hot continental division	1,670,000	1.14
M220 Hot continental regime mountains	485,000	0.33
230 Subtropical division	3,568,000	2.44
M230 Subtropical regime mountains	1,543,000	1.05
240 Marine division	1,347,000	0.92
M240 Marine regime mountains	2,194,000	1.50
250 Prairie division	4,419,000	3.02
M250 Prairie regime mountains	1,256,000	0.88
260 Mediterranean division	1,090,000	0.75
M260 Mediterranean regime mountains	1,561,000	1.07
300 Dry domain	46,806,000	32.00
310 Tropical/subtropical steppe division	9,838,000	6.73
M310 Tropical/subtropical steppe regime mountains	4,555,000	3.11
320 Tropical/subtropical desert division	17,267,000	11.80
M320 Tropical/subtropical desert regime mountains	3,199,000	2.19
330 Temperate steppe division	1,790,000	1.22
M330 Temperate steppe regime mountains	1,066,000	0.73
340 Temperate desert division	5,488,000	3.75
M349 Temperate desert regime mountains	613,000	0.42
400 Humid tropical domain	38,973,000	26.64
410 Savanna division	20,641,000	14.11
M410 Savanna regime mountains	4,488,000	3.07
420 Rainforest division	10,403,000	7.11
M420 Rainforest regime mountains	3,440,000	2.35

Source: World Conservation Monitoring Centre.

climate, we make use of climate diagrams of representative stations adapted from the well-known system of Heinrich Walter (Walter and Lieth 1960–1967; Walter et al. 1975). Soil information is founded on the soil classification established by the USDA Soil Conservation Service (U.S. Department of Agriculture 1938, Soil Survey Staff 1975), now the Natural Resources Conservation Service.

CHAPTER 5

Ecoregions of the Continents: The Polar Ecoregions

100 Polar Domain

Polar and arctic air masses chiefly control climates of the polar domain, located at high latitudes. With the exception of the icecap climates, they lie entirely in the Northern Hemisphere. In general, climates in the polar domain have low temperatures, severe winters, and small amounts of precipitation, most of which falls in summer. Polar systems are dominated by a periodic fluctuation of solar energy and temperature, in which the annual range is far greater than the diurnal range (see Figure 4.2, p. 35). This contrasts with the tropics, where the major periodic fluctuation is the diurnal one, and the mid-latitude systems, which we will see, are subject to fluctuations in both annual and diurnal energy patterns.

The intensity of the solar radiation is never very high compared to ecosystems of the middle latitudes and tropics. On the poleward margins, although summer insolation persists for many hours, temperatures do not get very high because the intensity is low and because much of the energy goes to evaporate water and melt snow or ice. More energy is given off by terrestrial radiation than is received from solar radiation. For this situation to persist, a supplementary heat source must exist to provide the difference. This supplementary energy source is heat carried poleward by wind and water currents. It maintains the arctic temperatures at a level much higher than would solar radiation alone.

The high-latitude climates have low annual total evaporation, always less than 50 cm, reflecting the prevailing low air and soil temperatures. The frozen condition of the soil in several consecutive winter months causes plant growth to cease. Snow that falls in this period is retained in surface storage until spring thaw releases it for infiltration and runoff. The growing season for crops is short in the subarctic zone, but low air and soil temperatures are partly compensated for by a great increase in day length.

In areas where summers are short and temperatures are generally low throughout the year, thermal efficiency, rather than effectiveness of precipitation, is the critical factor in plant distribution and soil development. Three major regional divisions have been recognized and delimited in terms of thermal efficiency: the *icecap*, *tundra*, and the *subarctic* (tayga). The world map of the polar ecoregions of the continents (Figure 5.1) shows the locations of these ecoregions. Climate diagrams in Figure 5.2 provide general information on the character of the climate in two of these divisions.

These climate diagrams express differences in the climatic regime of the divisions within the domains by comparing annual temperature and moisture cycles. A relatively dry season is depicted by the precipitation curve falling below the temperature curve. The location of the weather station and its altitude, as well as the average annual temperature and precipitation, are shown above the graphs.

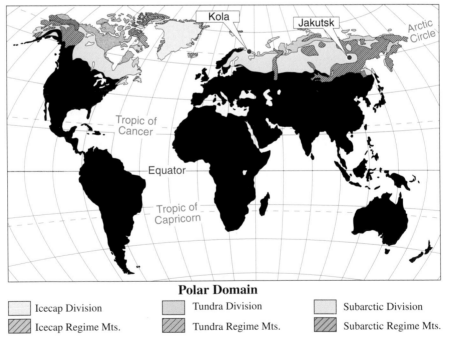

Polar Domain

Icecap Division	Tundra Division	Subarctic Division
Icecap Regime Mts.	Tundra Regime Mts.	Subarctic Regime Mts.

Figure 5.1. Divisions of the continental polar domain.

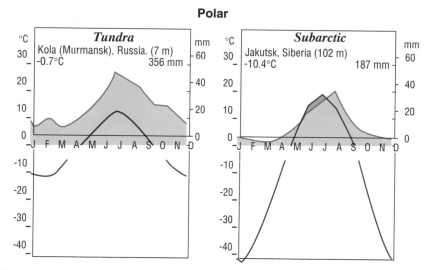

Figure 5.2. Climate diagrams from the tundra of northern Russia and from the extremely cold, continental boreal (tayga) regions of eastern Siberia. Redrawn from Walter et al. (1975).

110 Icecap Division

These are the ice sheets of Greenland, Ellesmere Island, Baffin Island, and Antarctica. Mean annual temperature is much lower than that of any other climate, with no month above freezing, defining this climate as *Fi*. Precipitation, almost all occurring as snow, is small, but accumulates because of the continuous cold. Driving blizzard winds are frequent. These regions are subjected to long periods of darkness and light.

Because of low monthly mean temperatures throughout the year over the ice sheets, this environment is almost devoid of vegetation and soils (Figure 5.3). Only a few plant species can survive the antarctic climate. The dominant plants are algae, lichens, and mosses—plants adapted to limited water supply and scant nutrients and soil. The few species of animals found on the ice margins are associated with a marine habitat. In the antarctic regions, because of the absence of predatory animals, the flightless penguins find conditions on the coasts and on the isolated islands of the southern oceans ideal as breeding places.

Terrestrial invertebrates include a few species such as nematodes and springtails. Some starve and dehydrate themselves during cold weather because any water left in their bodies would encourage deadly ice crystals to form. Others produce antifreeze chemicals.

Figure 5.3. A glacier and mountains in Antarctica. Taken during the Byrd Expedition to the South Pole, 1946–1947. Neg. no. 123435. Courtesy Department of Library Services, American Museum of Natural History.

120 Tundra Division

The northern continental fringes of North America, Iceland, Spitsbergen, coastal Greenland, and the arctic coast of Eurasia from the Arctic Circle northward to about the 75th parallel lie within the outer zone of control of arctic air masses. This produces the tundra climate that Trewartha (1968) designated by symbol *Ft*. Average temperature of the warmest month lies between 10°C and 0°C. The tundra regions occupy some 5% of the land surface of the earth.

The tundra climate has very short, cool summers and long, severe winters (see Figure 5.2, climate diagram for Kola [near Murmansk], Russia). No more than 188 days per year, and sometimes as few as 55, have a mean temperature higher than 0°C. Annual precipitation is light, often less than 200 mm, but because potential evaporation is also very low, the climate is humid. However, because precipitation is usually in the form of snow and ice, animals and plants cannot use it.

Tundra is the characteristic vegetation of the polar regions. Three chief types are recognized: these are *grass tundra*, *brush tundra*, and the *desert tundra*. Vegetation in the central part is grass tundra, a treeless plain of low-growing plants adapted to the climate's low temperatures, short growing season, and low precipitation. It consists of grasses, sedges, and lichens, with willow shrubs (Figure 5.4). As in the Antarctic, mosses and lichens flourish because they can tolerate the freezing temperatures. Farther south, the vegetation changes into brush tundra, or birch-lichen woodland, then into a needleleaf forest. In some places, a distinct tree line separates the forest from tundra. Köppen (1931) used this line, which coincides approximately with the 10°C

Figure 5.4. Grass tundra in Alaska. Photograph by USDA Forest Service.

isotherm of the warmest month, as a boundary between subarctic and tundra climates. Farther poleward the tundra breaks up into detached "oases" in the sheltered hollows, separated by expanses of bare rock or **regolith**. This is the desert tundra. The arctic flora is poor in plant species; only a few hundred species grow in the entire Arctic, compared to over 100,000 in the tropics.

In contrast to surfaces exposed to tropical rainy climate, soil particles of tundra derive almost entirely from mechanical breakup of the parent rock, by continual freezing and thawing, with little or no chemical alteration. **Tundra soils** (Entisols, Inceptisols, and associated Histosols), with weakly differentiated horizons, dominate. As in the northern continental interior, the tundra has a permanently frozen sublayer of soil known as **permafrost** (Figure 5.5). The permafrost layer is more than 300 m thick throughout the region; seasonal thaw reaches only 10 to 60 cm below the surface.

Probably half of the tundra surface is covered by water in the summer and ice in the winter. Poor drainage is a dominant characteristic. Permafrost prevents the percolation of meltwater into the regolith. The continental ice sheets that repeatedly scoured these areas left a rolling topography relatively free of weathered material. The depressions are filled with lakes and swamps, and bogs are everywhere. The streams that exist meander extensively on the surface, from depression to depression, or swamp to swamp. The large rivers that cross the tundra are **exotic rivers** and subject to spring flooding. Since these streams have their headwaters equatorward of the mouth, the spring thaw takes place first upstream. The meltwater starts downstream, only to encounter a channel blocked by a still-frozen stream. The only outlet for the meltwater is out of the channel over the ice and surrounding land. In this fashion, immense areas of the arctic tundra are flooded in the spring.

One aspect of the hydrologic cycle in the Arctic warrants particular notice. Whereas in most parts of the world the change of water to water vapor is a primary means of heating the atmosphere, in the Arctic this process is greatly reduced. Outside the Arctic, when condensation occurs and the water vapor is returned to a liquid or solid state, the energy utilized in evaporating water is again released into the atmosphere and heats it.

Geomorphic processes are distinctive in the tundra, resulting in a variety of curious landforms. Under a protective layer of sod, water in the soil melts in summer to produce a thick mud that sometimes flows downslope to create bulges, terraces, and lobes on hillsides. The freeze and thaw of water in the soil also sorts the coarse particles from the fine particles, giving rise to such patterns in the ground as rings, polygons, and stripes made of stone. The coastal plains have numerous lakes of **thermokarst** origin, formed by melting groundwater.

The richness and variety of the fauna of the arctic regions are re-

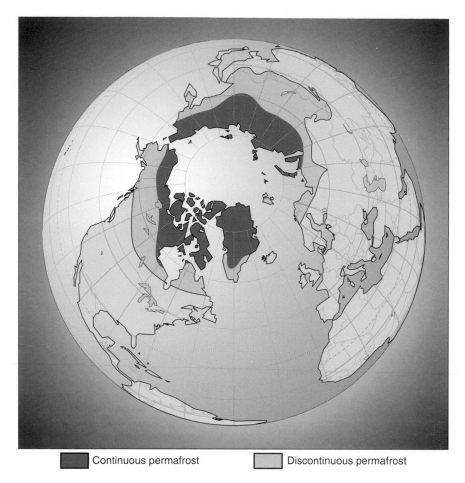

| Continuous permafrost | Discontinuous permafrost |

Figure 5.5. Distribution of permafrost in the Northern Hemisphere. From *Glacial and Quaternary Geology* by R.F. Flint, p. 270. Copyright © 1971 by John Wiley & Sons, Inc. Reprinted by permission of John Wiley & Sons, Inc.

markable. Many of the land animals of the tayga migrate northward onto the tundra during the summer months. The reindeer of Eurasia and the similar, but smaller, caribou of North America are the most important of these. Another large herbivore, the musk oxen, occurs in isolated herds, grazing on the more luxuriant patches of desert tundra. Two small herbivora—the Arctic hare and the lemming—are also widespread.

Following the herds are a number of predatory animals, such as wolf and fox. The drift ice of the polar sea or the immediate coastal margins is the habitat of the polar bear, preying chiefly on marine life, such as the seal and the walrus.

Birds and insects are particularly numerous. Mosquitoes are probably present in greater numbers in summer over the tundra than anywhere else on Earth. Flies are abundant, especially near human settlements. Attracted in part by this abundance of insect life, many migrating birds make the tundra their goal in summer. The boggy tundra offers an ideal summer environment for waterfowl, sandpipers, and plovers.

Sea life is rich. Although reptiles, amphibians, and fresh-water fish are absent, the inhabitants of the polar oceans are varied and numerous.

Direct human impact in the tundra regions has been minimal until recent years. Eskimo and Lapp cultures have lived as part of the ecosystem for thousands of years. Their numbers have been small, in keeping with the relatively low primary production of the tundra and adjacent seas. However, major disturbance to the permafrost terrain became evident in World War II, when military bases, airfields, and highways were hurriedly constructed without regard for maintenance of the natural protective surface. Recently, oil deposits were discovered on the north slope of Alaska, with oil exploration and drilling creating possibilities for environmental damage. Oil spills on the landscape and in coastal waters are attendant problems.

Succession in the Arctic is a very slow process, and the development has the potential for disturbing the local system so that the ecosystem will not recover. The equilibrium of the tundra is based on very low energy flow. This means little chemical weathering and slow rates of soil evolution and plant growth.

130 Subarctic Division

The source region for the continental polar air masses is south of the tundra zone between latitudes 50° and 70° N. The climate type here shows great seasonal range in temperature. Winters are severe, and the region's small amounts of annual precipitation are concentrated in the three warm months. This cold, snowy, forest climate, referred to in this volume as the boreal subarctic type, is classified as *E* in the Köppen–Trewartha system. This climate is moist all year, with cool, short summers (see Figure 5.2, climate diagram for Jakutsk, Siberia). Only one month of the year has an average temperature above 10°C.

Winter is the dominant season of the boreal subarctic climate. Because average monthly temperatures are subfreezing for six to seven consecutive months, all moisture in the soil and subsoil freezes solidly to depths of a few meters. Summer warmth is insufficient to thaw more than a meter or so at the surface, so permafrost and patterned ground prevails over large areas (Figure 5.6). Seasonal thaw penetrates from 0.5 to 4.0 m, depending on latitude, aspect, and kind of ground. De-

Figure 5.6. Patterned ground caused by alternating freezing and thawing of the ground overlying permafrost, northeast of Fort Yukon, Alaska. Photograph by T.G. Freeman, Soil Conservation Service.

spite the low temperatures and long winters, the valleys of interior Alaska and Siberia were not glaciated during the Pleistocene, probably because of insufficient precipitation (Figure 5.7).

The subarctic climate zone coincides with a great belt of needleleaf forest, often referred to as boreal forest, and open lichen woodland known as **tayga**. These species have adapted to the cold winter by greatly reducing their leaf area and by being able to respond rapidly to the short summer. Among the more widespread dominants are pine, fir, and spruce. Most trees are small, with more value to humans for pulpwood than for lumber. Different species occur in extremely wet and dry sites. In burned-over areas, a mixture of deciduous trees and evergreen is characteristic during secondary succession. In Siberia, this second-growth enclave of broadleaf types is called "white tayga." Slow growing conifers reproduce the climax forest only over a long period of time.

Figure 5.7. Extensive areas of Pleistocene glaciation are largely confined to the Northern Hemisphere. Adapted from *Glacial and Quaternary Geology* by R.F. Flint, p. 75. Copyright © 1971 by John Wiley & Sons, Inc. Reprinted by permission of John Wiley & Sons, Inc.

The forests run diagonally across the continents. On the west coast, the boreal forest is more than 10° farther north than on the east coasts. The contrast of warm and cold ocean water on the two sides of the continents in higher middle latitudes causes this diagonal arrangement.

The arctic needleleaf forest grows on **podzols** (Spodosols with pockets of wet, organic Histosols). The podsol profile is distinctly shallower than any other mature profile (see Figure 6.1, p. 64), in a few places reaching depths greater than 45 to 60 cm. Soil development is slow because the land is frozen for long periods each year. For various reasons, notably the absence of earthworms, the humus layer on the surface is not mixed with the soil, but remains as a very black, highly acidic accumulation. The lower part of the *A* horizon is strongly leached to a gray or even white color. A distinct layer of humus and forest litter lies beneath the top soil layer. The *B* horizon is reddish from the accumulation of part of the leached material and is very compact. Agriculture potential is poor, due to natural infertility of soils and the prevalence of swamps and lakes left by departed ice sheets. In some places, ice scoured the rock surfaces bare, entirely stripping off the overburden. Elsewhere rock basins were formed and stream courses dammed, creating countless lakes.

These lakes are only temporary features. Since decomposition is slow in the cold climate, these lakes gradually fill in with peat, organic matter produced by sphagnum moss or sedges, along with a definite succession of vegetation. These deposits have provided a low-grade fuel in northern Europe (Figure 5.8).

Peat is also an excellent insulator. In the far north, it keeps summer heat from completely thawing the frozen ground below the depth of 0.5 m or so. This subsoil, or permafrost, remains permanently frozen. Above it in the bog, annual freezing and thawing of the peaty soil

Figure 5.8. A peat bog in boreal forest near the border between Norway and Sweden. Photograph by John S. Shelton; reproduced with permission.

Figure 5.9. Tilting of a line of poles in a bog is also caused by freezing and thawing, Yukon region, Alaska. The cross-poles at the base are to minimize the effect. Photograph by T.L. Pewe, U.S. Geological Survey.

pushes against the trees because it cannot push the solid permafrost down. This annual freeze–thaw cycle produces a topsy-turvy forest with tree trunks and utility poles leaning in all directions (Figure 5.9).

The great north-flowing rivers of these regions, like those of the arctic tundra, are subject to extensive spring floods, and the lowlands, even where they are not glaciated, do not dry out rapidly after the water recedes. Permanently water-logged surfaces, in many cases underlain by peat, are common in all these regions. In Canada, the name "muskeg" is applied to such surfaces.

This is the habitat of large ground animals—a population that derives most of its food supply from the aquatic life of the numerous lakes, rivers, and swamps. In these forests roam the world's chief fur-bearing animals: minks, martens, foxes, wolves, badgers, bears, beavers, squirrels, sables, and ermines. There are several large ungulata, chief of which are deer, moose, the caribou, and reindeer.

CHAPTER 6

The Humid Temperate Ecoregions

200 Humid Temperate Domain

Both tropical and polar air masses govern the climate of the humid temperate domain, located in the mid latitudes (30° to 60°) on all the continents. The mid latitudes are subject to **cyclones**; much of the precipitation in this belt comes from rising moist air along fronts within those cyclones. Pronounced seasons are the rule, with strong annual cycles of temperature and precipitation. The seasonal fluctuation of solar energy and temperature is greater than the diurnal (see Figure 4.2, p. 35). The climates of the mid latitudes have a distinctive winter season, which tropical climates do not. The lower temperatures of the winter season are due to two factors: reduced solar radiation, and the inflow of cold air streams.

In the tropics (see Chapter 8), the different environments are distinguished on the basis of seasonal moisture pattern. There are regions with a high frequency of precipitation, and regions with strong seasonal contrasts. Although the tropics are subject to one annual periodicity, most humid temperate regions are subject to two major annual cycles, one of solar energy and another of moisture.

In the temperate latitudes, probably the most important aspect of the hydrologic cycle is the periodic freezing of lakes, streams, and soil moisture. This freezing stops the flow of water runoff. At the same time, it decreases the water supply available to the plants, producing drought. Most streams will reflect the winter season in the flow pattern. When the spring thaw occurs, flooding may occur if there are large amounts of snow in the watershed and melting takes place rapidly. Removal of headwater forests have increased flooding.

Many of the same process of soil development that operate in tropical forest lands are active here, but go on more slowly, owing to lower temperatures and less extreme humidity. Humus accumulates, too, for the slower decay of organic litter on the forest floor, when mixed with the soil layers, imparts a brownish color.

Three zonal soil types are recognized in the regions of this group (Figure 6.1). The red and yellow lateritic soils of the humid tropical group extend poleward into the warmer parts of the humid temperate lands and are known as **yellow forest soils** (Ultisols). Farther poleward, however, humus accumulation is sufficiently rapid so that soil color is darkened. With the aid of earthworms, the organic matter is mixed with the upper soil layers to form the **brown forest soils** (Alfisols). The podzols (Spodosols) lie on the northern borders of this group, extending into the polar group. In the profiles of the podzol, the absence of earthworms is indicated by the concentration of humus at the surface

Figure 6.1.
Generalized mature soil profiles developed under mid-latitude forests. After Jenny; from *A Geography of Man*, 2d ed., by Preston James, p. 197. Copyright © 1959 by Ginn and Company. Reprinted by permission of John Wiley & Sons, Inc.

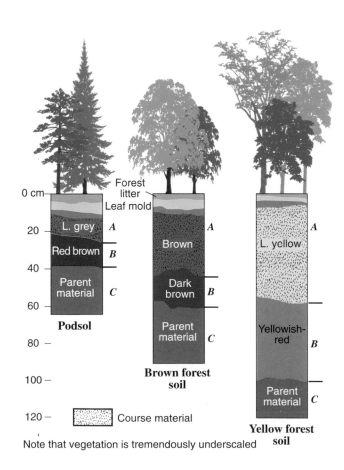

and the light, ashy color of the soil below (the name podzol is derived from a Russian word meaning ashes). The depth of these profiles decreases as the length of the frozen period of winter increases. None of these soils are fertile.

The regions of the humid temperate domain occur within climatic conditions where there is a winter cold season when plant growth ceases, and rain in summer is sufficient to support forest vegetation of broadleaf deciduous and needleleaf evergreen trees. On the equatorward side of these regions where this group borders the tropical regions, the differences in the forest are not sharply contrasted. Gradually the species that cannot survive frost drop out. On the poleward side of these regions in the Northern Hemisphere, there is a relatively sharp line of demarcation. The boreal forest occurs where there are severe winters and only short, cool summers. Here the spruce, fir, and larch are the more widespread conifers. Toward the continental interiors, on the dry side of these regions, grasslands usually border the forests.

Animals in these climates need to survive cold winters and seasonal variations in their food supply. Some vary their diet. Many birds and mammals migrate to warmer climates. Most amphibians and reptiles, as well as some large mammals, hibernate or become dormant.

The variable importance of winter frost determines six divisions: *warm continental*, *hot continental*, *subtropical*, *marine*, *prairie*, and *mediterranean*. Figure 6.2 shows the distribution of these divisions. Climate diagrams for these divisions are presented in Figure 6.3.

210 Warm Continental Division

South of the eastern area of the subarctic climate, between latitudes 40° and 55° N and from the continental interior to the east coast, lies the humid, warm-summer, continental climate. Located squarely between the source regions of polar continental air masses to the north and maritime or continental tropical air masses to the south, it is subject to strong seasonal contrasts in temperature as air masses push back and forth across the continent.

This climate occurs only in the Northern Hemisphere. It applies to the northeastern United States and southeastern Canada, southeastern Siberia, and northern Japan. Very similar climatic conditions also hold for eastern Europe across the Baltic countries and Russia as far as the Urals.

The Köppen–Trewartha system designates this area as *Dcb*, a cold, snowy, winter climate with a warm summer (see Figure 6.3, climate diagram for Moskva, Russia). This climate has four to seven months when temperatures exceed 10°C, with no dry season. The average tem-

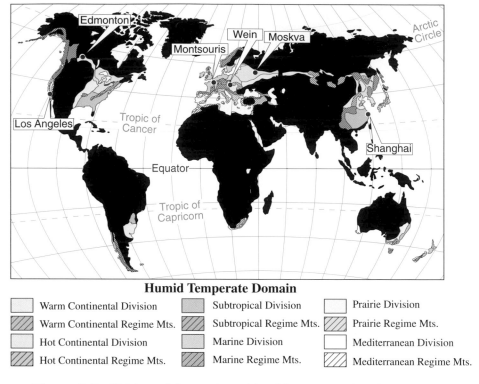

Humid Temperate Domain

Warm Continental Division	Subtropical Division	Prairie Division
Warm Continental Regime Mts.	Subtropical Regime Mts.	Prairie Regime Mts.
Hot Continental Division	Marine Division	Mediterranean Division
Hot Continental Regime Mts.	Marine Regime Mts.	Mediterranean Regime Mts.

Figure 6.2. Divisions of the continental humid temperate domain.

perature during the coldest month is below 0°C. The warm summer signified by the letter *b* has an average temperature during its warmest month that never exceeds 22°C. Precipitation is ample all year but substantially greater during the summer. In eastern Asia, a monsoon effect is strongly accentuated in summer.

Mixed boreal and deciduous forest grows throughout the colder northern parts of the humid continental climate zone (Figure 6.4) and is therefore transitional between the boreal forest to the north and the deciduous forest to the south. In eastern North America, part of it consists of mixed stands of a few coniferous species (mainly pine) and a few deciduous species (mainly birch, maple, and beech). The rest is a macromosaic of pure deciduous forest on favorable habitats with good soil and pure coniferous forest in less favorable habitats with poor soils. Here soils are gray-brown podsols (Alfisols). Such soils have a low supply of bases and a horizon in which organic matter and iron and aluminum have accumulated. They are strongly leached but have an upper layer of humus. Cool temperatures inhibit bacterial activity that

Figure 6.3. Climate diagrams from the various divisions of the humid temperate domain: mixed deciduous–coniferous forest regions (continental with cold winters and warm summers) and deciduous forest region (more moderate), broadleaf evergreen forest region (very rainy with hot summers), oceanic broadleaf forest region (rainy with warm summers), prairie region (cold winter with warm summer), and sclerophyllous regions of California (dry summer). Redrawn from Walter et al. (1975).

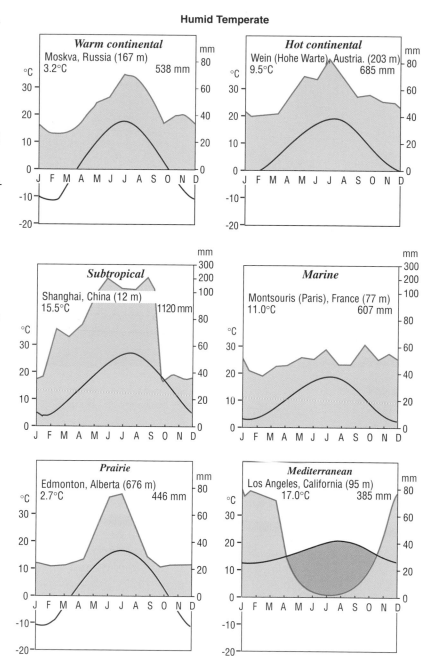

Humid Temperate

Warm continental
Moskva, Russia (167 m)
3.2°C   538 mm

Hot continental
Wein (Hohe Warte), Austria. (203 m)
9.5°C   685 mm

Subtropical
Shanghai, China (12 m)
15.5°C   1120 mm

Marine
Montsouris (Paris), France (77 m)
11.0°C   607 mm

Prairie
Edmonton, Alberta (676 m)
2.7°C   446 mm

Mediterranean
Los Angeles, California (95 m)
17.0°C   385 mm

Figure 6.4. Broadleaf deciduous and needleleaf evergreen forest as it appears in the Lake States region of north central United States. Photograph by Robert G. Bailey.

would destroy this organic matter in tropical regions. Deficient in calcium, potassium, and magnesium, soils are generally acidic. Thus, they are poorly suited to crop production, even though adequate rainfall is generally assured. Conifers thrive here.

Because of the availability of soil water through a warm summer growing season, this environment has an enormous potential for food production. North America and Europe support dairy farming on a large scale. A combination of acid soils and unfavorable glacial terrain in the form of bogs and lakes, rocky hills, and stony soils has deterred crop farming in many parts.

220 Hot Continental Division

South of the warm continental climate lies another division in the humid temperate domain, one with a humid, hot-summer continental climate. It has the same characteristics as the warm continental except that it is more moderate and has hot summers and cool winters (see

Figure 6.3, climate diagram for Wien, Austria). The boundary between the two is the isotherm of 22°C for the warmest month. In the warmer sections of the humid temperate domain, the frost-free or growing season continues for five to six months, in the colder sections only three to five months. Snow cover is deeper and lasts longer in the northerly areas.

This climate is located in central and eastern parts of the United States, northern China (including Manchuria), Korea, northern Japan, and central and eastern Europe.

In the Köppen–Trewartha system, areas in this division are classified as *Dca* (*a* signifies hot summer). We include in the hot continental division the northern part of Köppen's *Cf* (subtropical) climate region in the eastern United States. Köppen used as the boundary between the *C–D* climates the isotherm of −3°C for the coldest month. Thus, for example, in the United States, Köppen places New Haven, Connecticut and Cleveland, Ohio in the same climatic region as New Orleans, Louisiana and Tampa, Florida, despite obvious contrasts in January mean temperatures, soil groups, and natural vegetation between these northern and southern zones. Trewartha (1968) redefined the boundary between *C* and *D* climates as the isotherm of 0°C of the coldest month, thereby pushing the climate boundary south to a line extending roughly from St. Louis to New York City. Trewartha's boundary is adopted here to distinguish between humid continental and humid subtropical climates.

Natural vegetation in this climate is winter deciduous forest, dominated by tall broadleaf trees that provide a continuous dense canopy in summer but shed their leaves completely in the winter (Figure 6.5). Lower layers of small trees and shrubs are weakly developed. In spring, a luxuriant ground cover of herbs quickly develops but is greatly reduced after trees reach full foliage and shade the ground. Common trees in the eastern United States, eastern Europe, and eastern Asia are oak, beech, birch, hickory, walnut, maple, basswood, elm, ash, tulip, sweet chestnut, and hornbeam. Hemlock, a needleleaf evergreen tree, may also be present.

Soils are chiefly **red-yellow podzols** (Ultisols) and **gray-brown podzols** (Alfisols), rich in humus and moderately leached with a distinct, light-colored zone under the upper dark layer. The red-yellow podzols (Ultisols) have a low supply of bases and a horizon of accumulated clay. Where topography is favorable, diversified farming and dairying are the most successful agricultural practices. Because the gray-brown podzols (Alfisols) are soils of high base status, they proved highly productive for farming after the forests were cleared.

Most of the native forest has been cleared. The forest is preserved throughout the mountainous terrain of the Appalachian Mountains and

Figure 6.5. A stand of mature sugar maples in the Allegheny National Forest, Pennsylvania. Photograph by B.W. Muir, U.S. Forest Service.

the woodlots throughout the farmed belt. Much of this forest consists of second- or third-growth tree stands. Many farms in the forested area were abandoned and have since been covered by successional forest. The original animal life was abundant, with deer, bear, panthers, squirrels, and wild turkeys. Large mammals became scarce during the peak of agricultural use, but populations have increased with return of the forest environment.

In China and Korea, the effects of prolonged deforestation are evident everywhere. Throughout central Europe, large areas have been under field crops and pastures for centuries, while at the same time forests have been cultivated. Cereals grown extensively in North America and Europe include corn and wheat. In northern China, wheat is the principal crop. Rice is the dominant crop in both South Korea and Japan. Soybeans are intensively cultivated in the midwestern United States and in northern China and Manchuria, but very little in Europe.

230 Subtropical Division

The humid subtropical climate, marked by high humidity (especially in summer) and the absence of really cold winters, prevails on the eastern sides of the five continents in the lower middle latitudes, and is influenced by trade and monsoon winds. These areas include the southeastern United States, southern China, Taiwan (Formosa), southernmost Japan, Uruguay and adjoining parts of Brazil and Argentina, the eastern coast of Australia, and the North Island of New Zealand.

In the Köppen–Trewartha system, this area lies within the *Cf* climate, described as temperate and rainy with hot summers (see Figure 6.3, climate diagram for Shanghai, China). The *Cf* has no dry season; even the driest summer month receives at least 30 mm of rain. The average temperature of the warmest month is warmer than 22°C. Rainfall is ample all year but is markedly greater during summer. Rivers and streams flow copiously through much of the year. Thunderstorms, whether of thermal, squall-line, or cold-front origin, are especially frequent in summer. Tropical cyclones and hurricanes strike the coastal area occasionally, always bringing heavy rains and flooding. Winter fronts bring precipitation, some in the form of snow. Temperatures are moderately wide in range, comparable to those in tropical deserts, but without the extreme heat of a desert summer.

Soils of the moister, warmer parts of the humid subtropical regions are strongly leached red-yellow podzols (Ultisols) related to those of the humid tropical and equatorial climates. Rich in oxides of both iron and aluminum, these soils are poor in many of the plant nutrients essential for successful agricultural production. They are susceptible to severe erosion and gullying when exposed to forest removal and intensive cultivation.

Forest is the natural vegetation throughout most areas of this division. Along the outer coastal plain of the United States and in a large part of southern China and the south island of Japan, the native broadleaf forest was of the evergreen type. This forest consists of trees such the evergreen oak and trees of the laurel and magnolia families. Near its northern limits, vegetation of this region grades into broadleaf deciduous forest. Another type of rainforest is found in southeastern Australia and Tasmania and consists of many species of eucalyptus, which may reach heights of 100 m (Figure 6.6). The rainforest flora found in New Zealand consists of large tree ferns, large conifers such as the kauri tree, podocarp trees, and small-leaved southern beeches.

Figure 6.6. Eucalyptus forest in southeastern Australia. Photograph by Forests Commission of Victoria; from the American Geographical Society Collection, University of Wisconsin–Milwaukee Library.

Broadleaf evergreen forest may have a well-developed lower layer of vegetation that may include tree ferns, small palms, bamboos, shrubs, and herbaceous plants. Lianas and epiphytes are abundant.

Much of the southeastern United States today is covered by a second growth forest consisting of a number of pine species. It grows on the sandy soils of the coastal plain, where it appears to be a specialized type dependent on fast-draining sandy soils and frequent fires for its preservation.

Today, large areas have been converted to agricultural croplands, particularly in China. Corn is a major crop in the southern United States. Cattle production and tree farming are important uses of soils too sandy for field crops. Rice and tea are the most important crops in those parts of China and Japan with similar climate.

240 Marine Division

Situated chiefly on the continental west coasts and on islands of the higher middle latitudes between 40° and 60° N, is a zone that receives abundant rainfall from maritime polar air masses and has a narrow range of temperature because it borders on the ocean. These coasts and islands are bathed by warm ocean water, and the prevailing westerly winds bring abundant moisture to the land.

Trewartha (1968) classified the marine, west coast climate as *Do*— temperate and rainy, with warm summers. The average temperature of the warmest month is below 22°C, but at least four months of the year have an average temperature of 10°C. The average temperature during the coldest month of the year is above 0°C. Precipitation is abundant throughout the year but is markedly reduced during the summer (see Figure 6.3, climate diagram for Montsouris [Paris], France). Although total rainfall is not great by tropical standards, the cooler air temperatures reduce evaporation and produce a damp, humid climate with much cloud cover. Mild winters and relatively cool summers are typical. Coastal mountain ranges influence precipitation markedly in these middle latitudes. The mountainous coasts of British Columbia and Alaska annually receive 1,530 to 2,040 mm of precipitation and more. Heavy precipitation greatly contributed to the development of fiords along the coast in Norway (Figure 6.7), British Columbia, southern Chile, and the South Island of New Zealand. Heavy snows in the glacial period fed vigorous valley glaciers that descended to the seas, scouring deep troughs that reach below sea level at their lower ends. Farther back from the coast the annual rainfall decreases, even to less than 75 cm (London, 60 cm; Paris, 55 cm); but in the absence of any very high temperatures so little evaporation occurs that even this relatively small amount is highly effective and supports a luxuriant plant growth.

Needleleaf forest is the natural vegetation of the marine division. In the coastal ranges of the northwestern United States, it is the redwood zone. Farther north, this vegetation is succeeded by Douglas fir, western red cedar (Figure 6.8), and spruce which grow to enormous heights, forming some of the densest of all coniferous forest with some of the world's largest trees. Under the lower precipitation regime of Ireland, southern England, France, and the Low Countries, a broadleaf deciduous forest was the native vegetation. However, much of it disappeared many centuries ago under cultivation, so that only scattered forest plots or groves remain. Dominant tree species of this forest type in western Europe are oak and ash, with beech in the cooler, moister areas. In the Southern Hemisphere, the forests of Tasmania and New Zealand, and the mountainous coastal belt of southern Chile are of the **temperate rainforest** class.

Norge. Suldalsporten

Figure 6.7. This Norwegian fiord has the steep rock walls of a deep glacial trough. Postcard by Eneret Mittet and Co. [Place of publication unknown, author's collection]

The trees are covered with mosses, epiphytes, and ferns. They cover everything from living branches and leaves to rotting logs. Because of the cool climate, decomposition works slowly; logs pile upon logs to form a moss- and fern-covered jumble over the forest floor, which is deep in humus. This is very unlike the tropical rainforest, where decomposers make short work of falling debris so that the forest floor has little humus and few fallen logs.

Because of recent Pleistocene glaciation, landforms of glacial erosion and deposition are little changed from their original shapes. Glacial troughs and fiords are striking landforms. The extensive lowlands of northwestern Europe consist of till plains, moraines, and outwash plains left by ice sheets. Mountain watersheds, when disturbed by logging, can experience severe erosion and high sediment yields, particularly from mass wasting.

Soils of the marine regions bearing needleleaf forest are strongly leached, acidic brown forest soils (Alfisols). Due to the region's cool temperatures, bacterial activity is slower than in the warm tropics, so unconsumed vegetative matter forms a heavy surface deposit. Organic acids from decomposing vegetation react with soil compounds, re-

Figure 6.8.
Damp, oceanic coniferous forest with western red cedar in Mt. Baker National Forest, Washington. Photograph by E. Lindsay, U.S. Forest Service.

moving bases such as calcium, sodium, and potassium. Under the lower precipitation of the British Isles and western Europe, mid-latitude deciduous forests are underlain by gray-brown podzols (Alfisols).

Productivity is fairly high in these forests, but it chiefly goes into wood. Available forage is not high, so that the biomass of large animals is not great, although deer, elk, and mountain lions form part of the American marine ecosystem. In New Zealand, no mammals or reptiles existed in the original forest. The niches were filled by birds, such as the kiwi.

The marine regions of western Europe and the British Isles have been intensively developed for centuries for such diverse use as crop farming, dairying, orchards, and forests. In North America, the mountainous terrain is not suitable for agricultural use, except in limited valley floors. Forests are the primary resource here and constitute the greatest structural and pulpwood timber resource on Earth. Douglas fir, western cedar, and western hemlock are the principal lumber trees. The same mountainous terrain that limits agriculture is a producer of enormous water surpluses that run to the sea in rivers. These rivers support anadromous fisheries.

In spite of the heavy rainfall, periods of drought up to several weeks can occur, so that many areas in the northwestern United States and the western coast of New Zealand have been burned. In New Zealand, the introduction of many species of deer proved to be a serious mistake, since these forests evolved in the absence of grazing mammals. Deer and sheep have seriously overbrowsed these forests.

250 Prairie Division

The deciduous forests of the temperature zone are confined to climatic regions of an oceanic nature, where extremes of temperature are not sharp, and where rainfall is more or less evenly distributed throughout the year. Two chief kinds of grasslands thrive in the transition zone of the middle latitudes between the forests and the deserts. On the dry margins are the shortgrass **steppes** (p. 97) and on the wet margins are the tall grass **prairies**.

Prairies are typically associated with continental, mid-latitude climates designated as *subhumid*. Precipitation in these climates ranges from 510 to 1,020 mm per year and is almost entirely offset by evapotranspiration (see Figure 6.3, climate diagram for Edmonton, Alberta). In summer, air and soil temperatures are high. Soil moisture in the uplands is inadequate for tree growth, and deeper sources of water are beyond the reach of tree roots. In North America, prairies form a broad belt extending from Texas northward to southern Alberta and Saskatchewan. Similar prairies occur in the humid *pampa* of Argentina, in Uruguay and southern Brazil, in the *puszta* of Hungary and on the northern side of Russian steppes, in South Africa, in Manchuria, and in Australia.

In a transitional belt on the wetter border of the division, forest and prairie mix in a so-called **forest-steppe** or **parkland**. It is not a homogeneous vegetation formation like the tropical savanna (Chapter 8), but rather a macromosaic of deciduous-forest stands and prairie. Relief and soil texture determine the predominating vegetation. Forest are found on well-drained habitats, slightly raised ground, the sides of the river valleys, and porous soils; whereas the prairies occupy badly drained, flat sites with a relatively heavy soil (Figure 6.9). In the western prairies of the United States, the grassland has been changed with the introduction of trees. Forest were planted around the farmsteads and vil-

Figure 6.9. Prairie parkland on the Central Lowland, in northwest Iowa. Photograph by Robert G. Bailey.

lages, so that today the buildings are all but hidden in foliage during the summer.

In Canada and northern United States, the transition from grassland to boreal forest consists of a narrow belt of poplar and aspen deciduous forest. This belt is from 60 to 200 km wide.

The boundary between the prairie and the forest is not so clearly related to climatic or edaphic conditions. Prairies exist well within a climate humid enough for tree growth, and where trees are planted, they grow if protected from competition with the roots of the grasses. In previous times, fires caused by lightning and the grazing of big-game herds encouraged the growth of the grasses in the treeless, wet prairies.

The prairie climate is not designated as a separate variety in the Köppen–Trewartha system. Geographers' recognition of the prairie climate (Thornthwaite 1931, Borchert 1950) has been incorporated into the system presented here. Prairies lie on the arid western side of the humid continental climate, extending into the subtropical climate at lower latitudes. Temperature characteristics correspond to those of the adjacent humid climates, forming the basis for two types of prairies: temperate and subtropical.

Tall grasses associated with subdominant broadleaved herbs dominate prairie vegetation. The grasses of the Argentine Pampa are said to have once risen above the head of a man on horseback. Trees and shrubs are almost totally absent, but a few may grow as woodland patches in valleys and other depressions. Deeply rooted grasses form a continuous cover. They flower in spring and early summer, the forbs in late summer. In the tall-grass prairie of Iowa, for example, typical grasses are big bluestem and little bluestem; a typical forb is black-eyed Susan.

Because rain falls less in the grasslands than in forest, less leaching of the soil occurs. The pedogenic process associated with prairie vegetation is **calcification**, as carbonates accumulate in the lower layers. Soils of the prairies are prairie soils (Mollisols), which have black, friable, organic surface horizons and a high content of bases. Grass roots deeply penetrate these soils. Bases brought to the surface by plant growth are released on the surface and restored to the soil, perpetuating fertility. These soils are the most productive of the great soil groups.

These soils are not uniform and reflect the transitional nature of the climate. A succession of soil types, from the humid forest margins across the prairies and the steppes to the dry lands, conforms to the changes in moisture and vegetation cover (Figure 6.10). On the rainy margins of the prairie, a deep soil is formed which is so abundantly supplied with organic material that it is dark-colored even in the *B* horizon. This is the **black prairie soil** (Figure 6.11). Near the dry margin, rainfall decreases to the point that minerals dissolved near the surface are carried down to the *B* horizon and no farther.

Two soil types share this process of having mineral accumulations, chiefly lime, in the *B* horizon. The first of these, occupying the dry margins of the prairies, is known as the **chernozem** (Mollisol). The color of the chernozem is even darker than the black prairie soil, and its fertility is increased by the decreased effectiveness of the leaching process. The dry boundary of the chernozem coincides with the prairie–steppe boundary, where, because the depth of the moist surface soil becomes less than about 60 cm, the tall grasses give way to the short grasses. The smaller supply of humus from the short grasses is reflected in a change from the black color of the chernozem to a chestnut-brown color; and the more active evaporation and shallower penetration of the rain water result in the formation of a continuous layer of lime salts much closer to the surface than in the chernozem. This is the **chestnut-brown soil** (Mollisol).

Figure 6.10.
Generalized mature soil profiles that develop under mid-latitude grassland. After Jenny; from *A Geography of Man*, 2d ed., by Preston James, p. 312. Copyright © 1959 by Ginn and Company. Reprinted by permission of John Wiley & Sons, Inc.

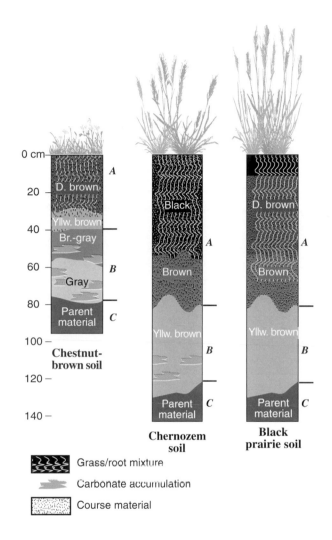

Most of the regions of this type are either plains or plateaus. The rainfall in the prairie is usually sufficient to support permanent streams, many of which are lined by **galeria forests**.

Most of the prairie has disappeared, replaced by some of the richest farmland in the world (Figure 6.12). Some of the native animals of the region enjoyed either an elimination of their natural enemies or an increase in the supply of food, which made possible a sudden and large increase in their numbers.

The agricultural region that developed on the central plains of the United States as a result of the spread of farm settlement on to the prairie is known as the Corn Belt. Here the practice of feeding maize

Figure 6.11. Corn planted in prairie soil, Iowa. Photograph by Robert G. Bailey.

to hogs and cattle and marketing the fattened animals is sustained by the vast areas of level land and the remarkably sustained fertility of the soils. Although maize is the principal crop worldwide, this region is also ideally suited to other crops, such as wheat, particularly in Russia. On the other hand, soybeans are intensively cultivated in the midwestern United States and in northern China and Manchuria.

260 Mediterranean Division

Situated on the western margins of the continents between latitudes 30° and 45° N is a zone subject to alternately wet and dry seasons, the transition zone between the dry west coast desert and the wet west coast. There are five such locations in the world. The largest borders the Mediterranean Sea. In North America, the area included in this division lies primarily in California. Other areas are found in Chile, in South Africa around Capetown, and in Australia, where it is divided into a western area around Perth and an eastern area around Adelaide. It occupies only 2% of the Earth's surface.

Figure 6.12. North American Corn Belt in a prairie region of South Dakota. Some patches of relict galeria forest follow the stream in the foreground. Photograph by John S. Shelton; reproduced with permission.

Trewartha (1968) classified the climate of these lands as *Cs*, signifying a temperate, rainy climate with dry, hot summers. The symbol *s* signifies a dry summer (see Figure 6.3, climate diagram for Los Angeles, California).

This climate is a product of subsidence associated with the subtropical high. In the summer, the high moves poleward over these areas, bringing essentially desert weather. In the winter, the anticyclonic circulation moves equatorward, allowing the westerlies to bring moisture into the area. These weather patterns often lead to summer wildfires (Figure 6.13).

The combination of wet winters with dry summers is unique among climate types and produces a distinctive natural vegetation of hardleaved evergreen trees and shrubs called **sclerophyll**, scrub woodland. This type of vegetation reduces water loss with leaves that are small, thick, and stiff, with hard, leathery, and shiny surfaces. Various forms of sclerophyll woodland and scrub are also typical. Trees and

Figure 6.13. Fire sweeping through California chaparral in summer. Photograph by Leonard F. De-Bano, U.S. Forest Service.

shrubs must withstand the severe summer drought—two to four rainless months—and severe evaporation. Although in the different continents different species compose the woodland, the appearance of the vegetation, resulting from its adaptation to the peculiarities of climate, is strikingly similar. The broadleaf evergreen woodlands of southern Europe, for example, are composed mostly of various kinds of oaks, whereas the similar forests of Australia are species of eucalyptus.

Because the winters are not cold enough, nor the summer droughts long enough to enforce a period of rest, there is no season when the leaves drop from the trees and growth ceases. In this way the mediterranean vegetation differs from selva (p. 112), which is evergreen but which has no seasonal rhythm. The woodland adapts itself to summer droughts. The trees are widely spaced and all plants have deep tap roots and a wide development of surface roots (Figure 6.14). The evaporation from the plants is diminished by a thick bark and by the sclerophyllous leaves.

Figure 6.14. Sclerophyll open woodland south of San Francisco, California. Most of the trees are oaks. Photograph by R.E. Wallace, U.S. Geological Survey.

In many areas, the original cover of woodland has been radically altered, probably by humans. At present, large areas of this group are covered by a thick, low growth of bushes and shrubs, known as *maquis* in Europe, *chaparral* in California, and *mallee* in southern Australia. Extreme flammability characterizes the chaparral during the long, dry summer. This poses an ever present threat to suburban housing which has expanded into chaparral-covered hillsides in California.

This has raised a problem of resource management in the suburban areas around cities. When an attempt is made to keep fires from starting, the brush grows thicker and accumulates a layer of debris underneath. As a result, the brush fires are more destructive than they were when smaller fires occurred each year. Furthermore, when a large fire has completely burned the cover, the torrential winter rains that follow produce mud flows and floods. The soil may be swept completely away, leaving only the bare rock exposed at the surface, while bordering valleys are filled with mud. This situation is worsened in southern California where soil becomes water repellent following fire.

After a fire, many mediterranean shrubs resprout from root crowns. The seeds of some species need fire to germinate and may lay dormant for years until the next fire.

Soils of this mediterranean climate are not susceptible to simple classification. Soils typical of semi-arid climates associated with grasslands are generally found. Severe and prolonged soil erosion following deforestation and overgrazing has left the mediterranean region with much exposed regolith and bedrock.

Animals survive fire by taking flight, or by retreating to underground burrows. Native mammals include deer, rabbits, and numerous rodents. In southern Europe and California, many of the native species have been replaced by large domestic grazers such as cows, sheep, and goats. Much of the birdlife is migratory, visiting mainly during the spring and fall. Resident birds tend to have short wings and long tails, an aid to maneuvering around shrubs.

This region is an important source of citrus fruits, grapes, and olives. In the Mediterranean, cork from the bark of the cork oak is also important. In central and southern California, citrus, grapes, avocadoes, nuts (almond, walnut), and deciduous fruits are extensively grown. Irrigated alluvial soils are also highly productive for vegetable crops such as carrots, lettuce, artichokes, strawberries, and forage crops (alfalfa).

This division lies closely hemmed in between high mountains and the sea. The surface features include small and isolated valley lowlands, bordered by hills and backed by high mountain ranges. The heavy rains in the highlands feed numerous torrential streams. The gravels and sands that they bring down with them to the lowlands are piled up in huge alluvial fans along the piedmonts. Delta plains grow where the rivers flow into the sea. Owing to the concentration of rain in the winter season, the regimen of these streams shows a maximum in that season. However, where the streams rise high enough in the mountains to reach the snow fields, the maximum flow comes during the melting period in the spring. The removal of forest from the mountains has seriously changed this regimen in many areas. The removal of the forest causes severe floods during the winter and spring, and during the summer these floods are followed by droughts, when the streams dry up. The original forest cover of most of the regions bordering the Mediterranean Sea has largely been removed, seriously affecting the habitability of the lowlands.

CHAPTER 7

The Dry Ecoregions

300 Dry Domain

The essential feature of a dry climate is that annual losses of water through evaporation at the Earth's surface exceed annual water gains from precipitation. Due to the resulting water deficiency, no permanent streams originate in dry climate zones. Because evaporation, which depends chiefly on temperature, varies greatly from one part of the Earth to another, no specific value for precipitation can be used as the boundary for all dry climates. For example, 610 mm of annual precipitation produces a humid climate and forest cover in cool northwestern Europe, but the same amount in the hot tropics produces semiarid conditions.

The tropical dry climates occupy the air masses in the subtropical high-pressure cells centered over the tropics of Cancer and Capricorn, both north and south of the equator, in the zone between 20° and 30°. This subsiding air mass is stable and dry. Dry land regions can be found on the western sides of all the continents. The dry lands also extend inland from the western sides of the continents, bending poleward in each hemisphere. The general regularity of the global pattern of dry climates is a reflection of the regular arrangement of certain climatic and water features. Figure 4.6 (p. 38) shows that cold water bathes parts of the west coasts of all the continents. Because evaporation is much less from cold water than warm water, the rainfall is much less here than continental margins bathed by warm water. The presence of dry lands along the east coast of South America in Patagonia is associated with a wide expanse of cold water of the South Atlantic Ocean (Falkland Island Current). In North America, the dry lands cannot extend so far toward the east as they do in Asia because of the movement of moist maritime air from the Gulf of Mexico up the Mississippi Valley versus the lack of such air moving across the Himalayas in Asia.

We commonly recognize two divisions of dry climates: the arid *desert* (*BW*), and the semi-arid *steppe* (*BS*). Generally, the steppe is a transitional belt surrounding the desert and separating it from the humid climates beyond. The boundary between arid and semi-arid climates is arbitrary, but it is commonly defined as one-half the amount of precipitation separating steppe from humid climates. These climates are displayed in Figure 7.1 and mapped in Figure 7.2.

Of all the climatic groups, dry climates are the most extensive; they occupy a fourth or more of the Earth's land surface (see Figure 4.7, p. 38). In these climates, many plants and animals have adapted to live with minimal rain, drying winds, and high temperatures.

Dry

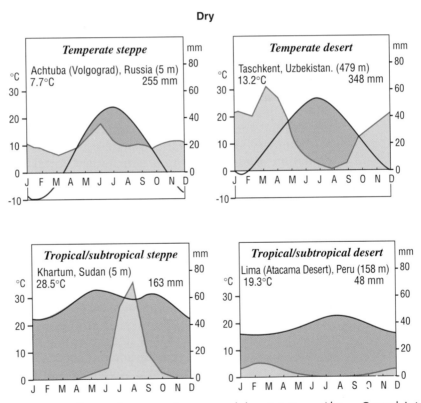

Figure 7.1. Climate diagrams of steppe and desert stations: *Above:* Central Asia, with some rain at all seasons, and winter rain. *Below:* With summer rain (northern Africa), and with rain that may fall at any season (Atacama Desert). Redrawn from Walter et al. (1975).

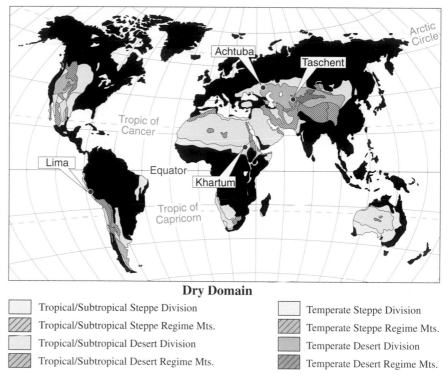

Dry Domain

Tropical/Subtropical Steppe Division		Temperate Steppe Division	
Tropical/Subtropical Steppe Regime Mts.		Temperate Steppe Regime Mts.	
Tropical/Subtropical Desert Division		Temperate Desert Division	
Tropical/Subtropical Desert Regime Mts.		Temperate Desert Regime Mts.	

Figure 7.2. Divisions of the continental dry domain.

310 Tropical/Subtropical Steppe Division

Tropical steppes occur along the less arid margins of the tropical deserts on both the north and south, and in places on the east as well. On the equatorward side of the deserts, it is transitional to the wet-dry tropical climate. Locally, altitude causes a semi arid steppe climate on plateaus and high plains that would otherwise be desert. Steppes on the poleward fringes of the tropical deserts grade into the mediterranean climate in many places. In the United States, they are cut off from the mediterranean climate by coastal mountains which allow the tropical deserts to extend farther north. Other important steppes of this type are the interior of the Kalahari Desert of South Africa, the dry eastern piedmont of the Andes, and northeastern Brazil.

Trewartha (1968) classified the climate of tropical/subtropical steppes as *BSh*, indicating a hot, semi-arid climate where potential evaporation exceeds precipitation and where all months have temperatures above 0°C (see Figure 7.1, climate diagram for Khartum, Sudan). Average rainfall is from 25 to 76 cm annually.

Steppes typically are grassland of short grasses and other herbs, with locally developed shrub and woodland. In the United States, pinyon–juniper woodland grows on the Colorado Plateau, for example. To the east, in New Mexico and Texas, the grasslands grade into savanna woodland or semi-deserts composed of xerophytic shrub and trees, and the climate becomes nearly arid-subtropical. Cactus plants are present in some places.

In the tropics, semi-desert is associated with this climate. A particularly important occurrence is the thorntree savanna of Africa, characterized by thorny trees and shrubs that shed their leaves for the long dry season. Another important area of this type is found in the Kalahari Desert of southern Africa, the home of the Bushmen. In the semi-desert zone of the African Sahel, the climate is associated with the acacia–desert grass savanna, where the stunted trees stand far apart and the short desert grasses cover most of the surface. In northwestern Brazil is an area once covered with thorny, scrubby deciduous trees known as *caatinga*. Where brush is thicker and mixed with trees it forms the scrub woodland of the Chaco in Argentina. Scrub woodland also forms a fringe around the desert in Australia (Figure 7.3).

Figure 7.3. Mulga scrub, a subtropical semi-desert, in western Australia. Neg. no. 335754 (Photo by R.A. Gould). Courtesy Department of Library Services, American Museum of Natural History.

Brown soils and sierozems (Mollisols, Aridisols) are associated with these tropical, semi-arid climates.

In Africa, nomadic herding is one of the main forms of agrarian uses, which is linked to the alternation of the dry and rainy seasons. During the dry season, the herders move into high-altitude regions, which are generally wetter, and during the rainy season, they move down to the lowlands again. In the deserts and semi-deserts, the herds consist of camels, sheep, and goats, while in the thorn savannas, cattle predominate.

Like the temperate steppe we describe below, rainfall in these regions can be expected to vary greatly and is subject to periods of drought interspersed with periods of ample rainfall. In several West African nations within the Sahelian zone, recent droughts have depleted grasses for grazing and devastated the annual grain crop. Some five million cattle perished, and many thousands of people died of starvation and disease. Periodic droughts in the past are well documented. In places, the land surface has been changed into desert through a process called **desertification**. Desertification in the African steppes can be attributed to greatly increased numbers of humans and their cattle.

320 Tropical/Subtropical Desert Division

The continental desert climates are south of the Arizona–New Mexico Mountains in the United States. They are not only extremely arid, but they also have extremely high air and soil temperatures. Direct solar radiation is very high, as is outgoing radiation at night, causing extreme variations between day and night temperatures and a rare nocturnal frost. Annual precipitation in deserts is less than 200 mm, and less than 100 mm in extreme deserts (see Figure 7.1, climate diagram for Lima, Peru). These areas have climates that Trewartha (1968) calls *BWh*.

The vast desert belt extending across North Africa (Sahara Desert), Arabia, and Iran to Pakistan (Thar Desert) are of this type. Other important deserts in this belt are the Sonoran Desert of the southwestern United States and northern Mexico and the Great Australian Desert.

Important climatic differences exist between interior and coastal deserts at the same latitude. Cold ocean currents upwelling from great depths lie offshore along the west coasts. The cold ocean current absorbs heat from the overlying air. Fog is a persistent feature and extends inland for short distances. Deserts here are cool and have low annual ranges of temperatures. The Atacama Desert of Chile and the Namib Desert of coastal southwest Africa are notable examples of these cool, foggy deserts.

Because desert rainfall is unreliable, river channels and the beds of smaller streams are dry most of the time (Figure 7.4). However, a sudden and intense downpour can cause local, brief flooding that transports large amounts of sediment. Major river channels, called **wadis** (or **arroyos** in the southwestern United States), often end in flat-floored basins having no outlet. Here clay and silt are deposited and accumulate, along with layers of soluble salts. Shallow salt lakes occupy some of these basins. Where the lakes are temporary, they are know as **playas**. In various low places in the hot desert, ground water can be reached by digging or drilling a well. Where such water supplies are available, they are often used to irrigate agricultural plots, creating an oasis (Figure 7.5).

Dry-desert vegetation characterizes the region. Widely dispersed xerophytic plants provide negligible ground cover. In dry periods, visible vegetation is limited to small hard-leaved or spiny shrubs, cacti, or hard grasses. Many species of small annuals may be present, but they appear only after the rare, but heavy, rains have saturated the soil.

Figure 7.4. The incised channel of the Rio Puerco, New Mexico, northwest of Albuquerque. It carries water only a few days each year. Photograph by Lev Ropes, Guru Graphics.

Figure 7.5. An oasis in the Sahara Desert of Libya. Date palms are planted in a sea of dune sand. Photograph by G.H. Goudarzi, U.S. Geological Survey.

In a variety of ways, the biota of the deserts have adapted to drought, a situation made worse by drying winds and high temperatures. Succulents resist drought by storing water inside their roots and stems, and protect themselves from evaporation by having a thick, waxy layer and no leaves. Their extensive surface root system quickly absorbs water before it sinks into porous soil. Some desert plants, such as the creosote bush, can survive without water, while others evade drought by growing close to a constant source of water, such as an oasis. Other plants, such as mesquite and tamarisk, send down very deep roots that are able to tap into a year-round supply of moisture. Some annual plants avoid the drought by lying dormant during the period between rains. Perennials, such ocotillo, become dormant between the rains. Once all moisture has evaporated from the soil, the plant drops its leaves and temporarily stops growing.

In the Mojave–Sonoran Deserts (American Desert) of the United States and Mexico, plants are often so large that some places have a near-woodland appearance (Figure 7.6). They include the tree-like saguaro cactus, the prickly pear cactus, the ocotillo, creosote bush, and smoke tree. However, much of the desert of the southwestern United States is in fact scrub, thorn scrub, savanna, or steppe grassland. Parts of these regions have no visible plants. They are made up of shifting dune sand, almost sterile salt flats, or bare mountain slopes covered by a mantle of loose rock fragments (Figure 7.7).

Over large areas of these deserts, the regolith has no soil. On the margins, however, where enough percolating water exists to moisten the upper layers of the regolith at more or less regular intervals, and where **xerophytes**, or xerophytic plants, are concentrated, a dry type of soil is formed; it is known as the **sierozem** (Aridisols). Humus is lacking, and this soil is gray in color at the surface, becoming lighter in the subsoil (Figure 7.8).

The dominant pedogenic process is **salinization**, which produces areas of salt crust where only salt-loving plants (halophytes) can survive.

Figure 7.6. Sonoran Desert near Tucson, Arizona. Photograph by Matthew G. Bailey.

Figure 7.7. Barren mountains in the Peruvian littoral desert at Santa Valley, north of Lima. Neg. no. 334565 (Photo by Shippee-Johnson). Courtesy Department of Library Services, American Museum of Natural History.

Figure 7.8. Creosote bush on light desert soil in southern Nevada. Photograph by Robert G. Bailey.

Calcification is conspicuous on well-drained uplands, where encrustations and deposits of calcium carbonate (caliche) are common.

Because there is little water in the deserts, mechanical weathering, or physical disintegration of the bedrock, is more rapid than chemical weathering. When a rock, such as granite, is composed of minerals of different colors, each mineral expands and contracts at a different rate. Such rocks quickly crumble into coarse sand, which forms an abrasive agent when picked up by the wind (Figure 7.9). The very rough and youthful appearance of the desert topography is largely the result of this process. There are three kinds of desert landform regions: the **erg**, or sandy desert (Figure 7.10), the **hamada** deserts composed of rocky plateaus channeled by dry water courses (see Frontispiece), and basin-and-range deserts of basins surrounded by barren mountains.

Figure 7.9. Granite hollowed out by windblown sand, Atacama Desert, Chile. Photograph by K. Segerstrom, U.S. Geological Survey.

Figure 7.10. An erg landscape in the Gobi Desert. Taken during the Roy Chapman Andrews Third Asiatic Expedition, 1925. Neg. no. 315830 (Photo by J.B. Shackelford). Courtesy Department of Library Services, American Museum of Natural History.

Animals have adapted to the desert environment. Some very specialized forms, of which the classic example is the camel, must have fresh water but are able to store considerable quantities. Other species, such as scorpions, are nocturnal, carrying on their life activities at night when it is cooler and less water is needed for temperature control. Snakes and tortoises retreat to burrows. Still others, such as kangaroo rats of the American desert, restrict their water needs to what water they can manufacture by metabolism. These animals usually have little evaporation loss from the surface and extremely dry waste products.

The world's hot deserts have been productive. Wherever irrigation is possible, the yields of crops have been high. The crops include long-staple cotton, fruits (dates, oranges, lemons, grapefruit, limes), vegetables, grains, and alfalfa. However, such irrigation projects suffer from two undesirable side effects—salinization and waterlogging of the soil. If irrigation water is evaporated in the soil it leaves salts behind. When these are sodium salts, and they accumulate too rapidly, they form an impervious **alkali**. When drainage is not adequate to flush out the excess sodium salts, this alkali may render desert soil completely sterile.

Like the arctic tundra, the processes of soil development and plant succession are highly vulnerable to change from human activities. A single passage of an army tank over desert soil can dramatically alter water infiltration, soil moisture, and heat distribution, and consequently, biological productivity. For example, in the Mojave Desert, lichen crust on sandy soil destroyed in tracks left by tank traffic in World War II-vintage maneuvers has not recovered in fifty years.

Figure 7.11. A dust storm approaching in the steppe of eastern Colorado. Photograph by Soil Conservation Service.

330 Temperate Steppe Division

Temperate steppes are areas that have a semi-arid continental climatic regime in which, despite maximum summer rainfall, evaporation usually exceeds precipitation. There is too little water to support a forest, and too much to create a desert. Instead these regions are dominated by grasslands, which are called *shortgrass prairie* in the central United States (Columbia Plateau, Great Plains), *steppe* in Eurasia, *pampas* in South America, and *veldt* in Africa. With more moisture, the vegetation changes to savannas, which are grasslands with enough moisture to support sparse tree growth.

Trewartha (1968) classified the climate as *BSk*. The letter *k* signifies a cool climate with at least one month of average temperature below 0°C. Winters are cold and dry, summers warm to hot (see Figure 7.1, climate diagram for Achtuba [Volograd], Russia). Drought periods are common in this climate. With the droughts come the dust storms that blow the fertile topsoil from vast areas of plowed land being used for dry farming (Figure 7.11).

These regions are subject to periodic climatic shifts. Thus, the drought typical of arid zones may extend well outside the normal desert boundaries one year, and precipitation characteristic of wetter regions may make incursions into an arid zone in the next year (Figure 7.12).

Figure 7.12. Climatic variations in the Great Plains of the United States during normal times and a drought in 1934. From Thornthwaite (1941), p. 182.

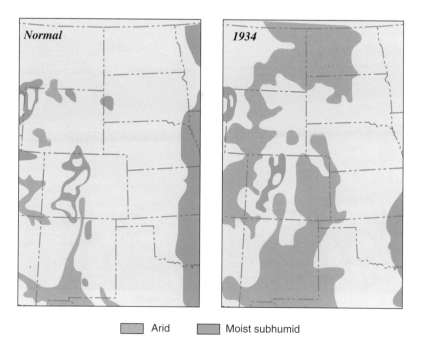

Arid   Moist subhumid

lands is found underground. Rodents—such as the prairie dog in America and hamsters in Eurasia—retreat underground to escape predators and the summer heat.

These steppes constitute the great sheep and cattle ranges of the world. The steppes of central Asia have for centuries supported nomadic populations. Wheat and, to a lesser extent, oats, rye, and barley, are dominant crops. The North American Great Plains, the Ukraine, and parts of north China are all within this region. Large areas of steppe have disappeared, replaced by farmlands (Figure 7.15).

In the Occidental world, farmers cultivate poorer land, supporting themselves on large areas by using machinery. On the dry margins they have managed to survive by allowing the land to lie fallow for a year, so that enough moisture might be stored up to permit crop production the following year.

340 Temperate Desert Division

Temperate deserts are located in the interior of continents, although they merge with tropical deserts equatorward. They are found in North

Figure 7.15. Plowing with steam tractor in the steppe of North Dakota during the early 1900s. Postcard by Bloom Bros., Minneapolis, author's collection.

America in the Great Basin, from Arizona northward into Washington. In Eurasia they are found embedded in the trans-Eurasian cordillera or lie on the flanks of these mountains. Most occur in the Turkestan and Gobi regions of central Asia. In the Southern Hemisphere, only South America projects far enough into mid latitudes to have a temperate desert climate, and only on the east side of the Andes Mountains in Patagonia, Argentina.

These desert regions have low rainfall and strong temperature contrasts between summer and winter. In the intermountain region of the western United States, between the Pacific and Rocky Mountains, the temperate desert has characteristics of a sagebrush semi-desert, with a pronounced drought season and a short humid season. Most precipitation falls in the winter, despite a peak in May (see Figure 7.1, climate diagram of Taschkent, Uzbekistan). Aridity increases markedly in the rain shadow of the Pacific mountain ranges. Even at intermediate elevations, winters are long and cold, with temperatures below 0°C.

These desert areas have the highest percentage of possible sunshine of any of the mid-latitude climates. Because of low humidity, over 90% of the sun's radiation reaches the ground. These deserts experience huge daily temperature fluctuations. During the summer when the days are long and the sun high in the sky, temperatures will reach 50°C. At sunset, heat is lost rapidly because of the lack of insulating clouds. The nighttime temperature can drop over 44°C from the daytime high, with winter temperatures in Asia's Gobi Desert plummeting to −21°C.

Under the Köppen–Trewartha system, this is the true desert, *BWk*. The letter *k* signifies that at least one month has an average temperature below 0°C. These deserts differ from those at lower latitude chiefly in their far greater annual temperature range and much lower winter temperatures. Unlike the dry climates of the tropics, middle-latitude dry climates receive a portion of their precipitation as snow.

Temperate deserts support the sparse xerophytic shrub vegetation typical of semi-deserts. One example is the sagebrush vegetation of the Great Basin (Figure 7.16) and northern Colorado Plateau region of the United States. Recently, semi-desert shrub vegetation seems to have invaded wide areas of the western United States that were formerly steppe grasslands, due to overgrazing and trampling by livestock. Soils of the temperate desert are sierozems (Aridisols), low in humus and high in calcium carbonate. Poorly drained areas develop saline soils, and salt deposits cover dry lake beds.

Productivity is low due to drought and low temperatures. Growth is thus limited to a short period between winter cold and summer drought. Much of the small production goes into woody tissue. This is in contrast to the production of digestible foods in grasslands. Consequently, although there are herbivores in the desert, the weight in animals per unit area is small.

Figure 7.16. Temperate semi-desert in Wyoming. Postcard by Sanborn Souvenir Co., Denver, author's collection.

Figure 7.17. Desert vegetation and dissected alluvial fan in Mongolia. Taken during the Roy Chapman Andrews Third Asiatic Expedition, 1925. Neg. no. 322017 (Photo by J.B. Shakelford). Courtesy Department of Library Services, American Museum of Natural History.

The surface features of these deserts consist of vast depressions surrounded by mountains or basin-and-range where flat-floored **bolsons** separate irregularly placed desert ranges. Alluvial fans surround the mountains ranges (Figure 7.17). Water from the mountains is abundant and provides support of large oasis communities at the base of the mountains and along the courses of exotic rivers.

CHAPTER 8

The Humid Tropical and Mountain Ecoregions

400 Humid Tropical Domain

Equatorial and tropical air masses largely control the humid tropical group of climates found at low latitudes. Every month of the year has an average temperature above 18°C, and no winter season. In these tropical systems, the primary periodic energy flux is diurnal: the temperature variation from day to night is greater than from season to season (see Figure 4.2, p. 35). Average annual rainfall is heavy and exceeds annual evaporation, but varies in amount, season, and distribution.

Two types of climates are differentiated on the basis of the seasonal distribution of precipitation. Figure 8.1 shows the global distribution of these two types, and Figure 8.2 is the climographs for two stations in humid tropical climates. *Tropical wet* (or rainforest) climate has ample rainfall through ten or more months of the year. *Tropical wet-and-dry*, or savanna, climate has a dry season more than two months long.

The circulation of the atmosphere controls the temporal pattern of precipitation in the tropics. Near the equator the trade winds of both hemispheres converge to form a low-pressure trough with a gentle upward drift of air. As mentioned before, this convergence zone is often referred to as the intertropical convergence zone, or ITC (see Figure A.1, p. 136), because it represents the zone in which the trade winds from the north and south of the equator converge. Where the converging air has a trajectory over the ocean, it contains large amounts of moisture, and cloudiness and frequent precipitation are common. Daily thunderstorms and torrential downpours predominate. Poleward of the zone, the air subsides near the tropics of Cancer and Capricorn and

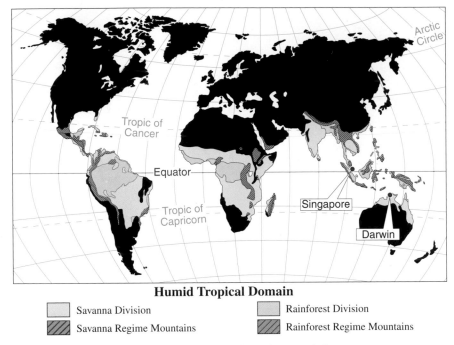

Humid Tropical Domain

☐ Savanna Division	☐ Rainforest Division
▨ Savanna Regime Mountains	▨ Rainforest Regime Mountains

Figure 8.1. Divisions of the continental humid tropical domain.

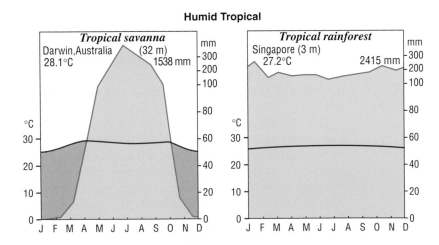

Figure 8.2. Climate diagrams of savanna and rainforest stations: (left) with maximum rain during the high-sun period; (right) with constantly wet climate. Redrawn from Walter et al. (1975).

aridity results. The ITC shifts north and south following the migration of the vertical rays of the solar energy. This migration produces the seasonal pattern of precipitation that characterizes so much of the tropical region.

The soils of the tropics include many large areas that cannot sustain continued crop cultivation, as well as mid-latitude soils. Temperature and moisture availability are high, with rapid chemical weathering. The weathered regolith is thus quite deep. Soil profiles are often 3 m or more, and evidence of chemical weathering has been found as deep as 70 m.

Under these conditions, a process of soil development called **laterization** takes place. In the process, iron, aluminum, and manganese form soluble hydroxides which tend to concentrate in the topsoil (Figure 8.3). Highly enriched layers of iron and aluminum hydroxides,

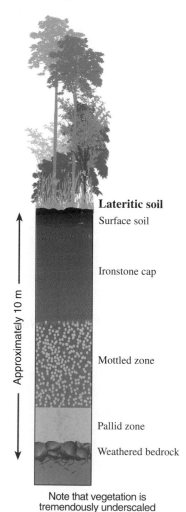

Figure 8.3.
Generalized profile of soil that develops in tropical climatic regimes. From King (1967), p. 175.

Lateritic soil
Surface soil

Ironstone cap

Mottled zone

Pallid zone

Weathered bedrock

Approximately 10 m

Note that vegetation is
tremendously underscaled

known as **laterite**, stain the soils reddish. Due to rapid chemical decomposition and solution, the soils are low in mineral nutrients. The topsoil contains little of the essential elements for plant growth. The soils are also low in humus, because litter decomposes quickly. Without fertilizers, these soils can sustain crops on freshly cleared areas for only two or three years before the nutrients are exhausted and the plot abandoned. This kind of migratory agriculture, although much reduced in areal distribution, is still one of the most important of the world's agricultural systems.

410 Savanna Division

The latitude belt between 10° and 30° N is intermediate between the equatorial and middle-latitude climates. This produces the tropical wet-dry savanna climate, which has a wet season controlled by moist, warm, maritime, tropical air masses at times of high sun, and a dry season controlled by the continental tropical masses at times of low sun (see Figure 8.2, diagram for Darwin, Australia). Trewartha (1968) classified the tropical wet-dry climate as *Aw*, the letter *w* signifying a dry winter.

Figure 8.4. Scrub woodland near Mysore in southern India. Photograph by John S. Shelton; reproduced with permission.

Alternating wet and dry seasons result in the growth of three distinctive vegetation types. In some areas no grass grows at all; in others only grass grows; and in other large areas grass and trees intermingle.

The tropical scrub woodlands are made up of trees standing far enough apart so the crown of foliage fails to form a complete cover. The tree trunks are gnarled with branches all the way to the ground. The trees are deciduous, dropping all their leaves in the dry season. There are areas of low brush, called *monte* in Argentina. Where the brush is thicker and mixed with trees it forms the scrub woodland of the Chaco. In northeastern Brazil is an area covered by thorny, scrubby deciduous trees, known as the *caatinga*. Similar scrub woodlands are found in Central America and Mexico. There are also large areas in Africa, Angola, Zambia, and Zimbabwe, and also in India in the dry parts of the Deccan (Figure 8.4). It also forms a fringe along the northern coast of Australia.

Most of the area of this type is covered with woodland savanna (Figure 8.5). It is characterized by open expanses of tall grasses, inter-

Figure 8.5. African savanna in Kenya. The trees are acacia. Neg. no. 211081 (Photo by Carl Akeley). Courtesy Department of Library Services, American Museum of Natural History.

spersed with hardy, drought-resistant shrubs and trees. Some areas have savanna woodland, monsoon forest, thornbush, and tropical scrub. In the dry season, grasses wither into straw, and many tree species shed their leaves. Other trees and shrubs have thorns and small or hard, leathery leaves that resist water loss. The major areas of occurrence are in Africa, covering a large part of the low latitudes, and in the interior of Brazil.

Ribbons of dense tropical forest along most of the streams are a distinguishing characteristic of this type of region. Where the streams are small, the tall trees form a complete arch so that the streams flow through tunnels, described as galeria forests.

Plains and plateaus are the surface features most commonly found in the regions of savannas (Figure 8.6). Some hilly uplands exist but almost no low mountain areas. Low mountains and, to a certain, extent, hilly uplands receive more rain than nearby plains or plateaus,

Figure 8.6. The flat, featureless plain of the Orinoco Llanos in Venezuela is covered with a mixture of savanna and scrub woodland. Drawing by Susan Strawn, from a photograph.

and where sufficient rain falls, the seasonal droughts are neither too long nor too dry, so forest can survive.

Soils are mostly **latosols** (Oxisols). Heavy rainfall and high temperatures cause heavy leaching. Streamflow in these regions is subject to strong seasonal fluctuations, in striking contrast to the constant streamflow typical of rainforest climates. In the rainy season, extensive low-lying areas are submerged; in the dry season, streamflow dissipates, exposing channel bottoms of sand and gravel as stream channels and mud flats dry out.

Animal life is rich in these regions. The greatest population of herbivores and carnivores of any region on earth is found in the African zone. Africa, in particular, is so noted for its variety of species that it has been the big game-hunting center of the world.

The dry season brings a severe struggle for existence to animals of the African savanna. As streams and hollows dry up, the few muddy waterholes must supply all drinking water (Figure 8.7). Danger of attack by carnivores is greatly increased.

Figure 8.7. Zebras and gnus at waterhole in Africa. Neg. no. 314034 (Photo by Martin Johnson). Courtesy Department Library Services, American Museum of Natural History.

420 Rainforest Division

The wet equatorial or rainforest climate lies between the equator and latitude 10° N. Average annual temperatures are close to 27°C; seasonal variation is imperceptible. Rainfall is heavy throughout the year, but the monthly averages differ considerably due to seasonal shifts in the ITC, and a consequent variation in air-mass characteristics (see Figure 8.2, climate diagram for Singapore). Trewartha (1968) defines this climate as *Ar*, with no month averaging less than 60 mm of rainfall.

The equatorial region has a forest made up of trees that cannot survive low temperatures. There are two kinds: in very wet areas with a dry season, tropical rainforest grows; where a distinct dry season occurs, tropical deciduous forest grows.

The rainforest, or **selva** type of vegetation, is unsurpassed in number of species and luxuriant tree growth (Figure 8.8). Broadleaf trees

Figure 8.8. Characteristic buttressing at base of cedro espinosa tree of the rainforest on Barro Colorado Island, Canal Zone. Neg. no. 2A4242. Courtesy Department of Library Services, American Museum of Natural History.

rise 30 to 45 m, forming a dense leaf canopy through which little sunlight can reach the ground (Figure 8.9). Giant lianas (woody vines) hang from trees. The forest is mostly evergreen, but individual species have various leaf-shedding cycles.

Where the moisture supply is less abundant, or where a dry season imposes a partial rhythm, a semideciduous, lighter forest grows. The dry season imposes a period of rest on the winter vegetation, just as winter imposes a similar period on the forests of the middle latitudes. Many, although not all, of the trees lose their leaves. On the dry margins, the semideciduous forest merges into woodlands.

Where both the rainforest and semideciduous forest border the ocean along low, coastal plains, mangrove forests develop (Figure 8.10). True mangrove consists of only one kind of tree of the genus *Rhizophora*. They grow on the shores of tidal swamps where the water is brackish and consists of a dense tangle of evergreen trees which grow some 5 to 7 m in height, with spreading bushy branches and numerous aerial roots. They serve as essential nurseries for many fish. They have been extensively cut in tropical lands to permit the entrance of ships, for making charcoal, and for aquaculture.

Tropical forests are home to small forest animals able to live and travel in the continuous forest canopy. These include insects, reptiles, cats, lemurs, and monkeys. Bird species are numerous and spectacularly plumaged. Underneath, in the deep shadows, are a multitude of insects, such as ants and spiders. Termites are particularly destructive of organic matter that falls to the ground. On the forest floor, there are larger mammals, including both herbivores, which graze the low leaves and fallen fruit, and carnivores, which prey on the herbivores. In Africa, there are elephant, buffalo, hippopotamus, okapi, bongo, and crocodile. In the South American forest lives the anteater, which thrives on termites and ants of all kinds.

A major difference between the low-latitude rainforests and forests of higher latitude is the great diversity of species. The tropical rainforest may have as many as 1150 different tree species in a square kilometer. The fauna of the rainforest is also very rich. A 16-km^2 area in the Canal Zone, for example, contains about 20,000 species of insects, whereas in all of France only a few hundred exist.

Copious rainfall and high temperatures combine to keep chemical processes continuous on the rocks and soils. Leaching of all soluble elements of the deeply decayed rock produces red and yellow podzols (Ultisols) and latosols (Oxisols) that are often especially rich in hydroxides of iron, magnesium, and aluminum.

Streamflow is fairly constant because the large annual water surplus provides ample runoff. Dense vegetation lines river channels. Sand bars and sand banks are less conspicuous than in drier regions. Floodplains have cutoff meanders (oxbows) and many swampy sloughs where me-

Figure 8.9. Rainforest canopy in Papua New Guinea. Photograph by H. Gyde Lund, U.S. Forest Service.

Figure 8.10. Mangrove forest in Everglades National Park, Florida. Photograph by M.W. Williams, United States Department of the Interior National Park Service.

andering river channels have shifted their courses. Although water is abundant, river systems carry relatively little dissolved material because thorough leaching of soils removes most soluble mineral matter before it reaches streams.

Not all equatorial rainforest areas have low relief. Hilly or mountainous belts have very steep slopes and frequent earthflows and slides; avalanches of soil and rock strip surfaces down to bedrock.

Man is steadily encroaching on the rainforest with lumbering, clearing for plantations, and other kinds of agriculture (Figure 8.11). As a result, the rainforest in many places has been eliminated or decimated and replaced by scrub or open savanna, which are subject to annual fires.

Figure 8.11. Lumbering on the Island of Mindoro, Philippines. Postcard by L.J. Lambert, Manila, author's collection.

M Mountains with Altitudinal Zonation

Highland systems are characterized by change more than any other regional system. On the geologic time scale they are subject to rapid changes in topography. Highland areas are associated with the margins of crustal plates, and the great altitudes result from the upwarping of the crust along the plate boundaries and the upwelling of magma that forms the volcanic peaks and massive lava flows. These are the zones where volcanic activity is common, and where earthquakes may be expected. The high relief, steep slopes, and generally higher precipitation accelerate erosional processes. Mass wasting is a widespread phenomenon in the highlands, including avalanches and landslides (Figure 8.12). It is these relatively high rates of geologic modification by both internal and external forces that give the highlands their rugged aspect.

Mountain climates are vertically differentiated, based on the effects of change in altitude. Air cools while ascending a mountain slope, and its capacity to hold water decreases, causing an increase in rain and snow. The thin, dry air loses heat rapidly as it ascends, and after sunset, temperatures plummet.

Figure 8.12.
Landslides that produce a variation in landform are widespread in highlands. From W.M. Davis, The lakes of California, *California Journal of Mines and Geology* 29, p. 202, 1933.

Elevations show typical climatic characteristics, depending on a mountain range's location in the overall pattern of global climatic zones. The climate of a given highland area is usually closely related to the climate of the surrounding lowland in seasonal character, particularly the form of the annual temperature cycle, and the times of occurrence of wet and dry seasons. For example, the Ethiopian Plateau is subject to a diurnal energy pattern, a nonseasonal energy pattern, and a seasonal moisture regime consisting of a rainy summer and a dry winter (Figure 8.13).

Since high mountains extend vertically through several climatic zones, their vegetation is usually rather sharply marked into zones. The altitudinal limits of various types of vegetation also correspond to the pattern of vertical temperature distribution. Roughly the same succession of types is found ascending mountains near the equator as is found proceeding poleward along the continental east coasts. As discussed in Chapter 4, each mountain within a zone has a typical se-

Figure 8.13.
Climate diagram of Addis Abeba, 2440 m above sea level, a tropical savanna regime highlands. Data from Walter et al. (1975).

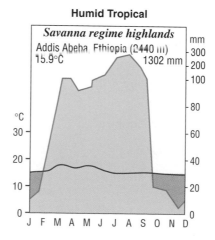

Humid Tropical

Savanna regime highlands
Addis Abeba, Ethiopia (2440 m)
15.9°C   1302 mm

quence or spectra of altitudinal belts (Table 8.1). For example, in the tropical rainforest, the tropical forest occupies the lower slopes; higher up are the mixed montane forests of broadleaf types with epiphytes (Figure 8.14); and beyond the upper limit of trees but below the snow line is a zone of alpine meadows. Conifers do not appear at higher altitudes south of Nicaragua.

As far as vegetation and other forms of life are concerned, the vertical differentiation reaches a maximum in the low latitudes. Here we find the greatest variety of zones. Vertical differentiation remains a conspicuous feature of mountains in the middle latitudes, but disappears altogether in the high latitudes.

The position of the montane zone varies with latitude. It starts at 2700 m in the Himalayas, 1200 m in the Sierra Nevada of California, 900 m in the western Alps of Europe, and sea level in the Chugach Mountains of Alaska. Long, cold winters and heavy snowfall create ideal conditions for evergreen conifers, such as pines, fir, and spruce (Figure 8.15). Food is limited in coniferous forests, and mountain animals, such as deer and many birds, migrate to higher elevations as the weather warms and food becomes abundant. In the fall, they reverse the migration. Animals not adapted to living in the cold mountain win-

Table 8.1. List of the types of altitudinal spectra

Name of division	Altitudinal spectra
110 Icecap	Polar desert
120 Tundra	Tundra — polar desert
130 Subarctic	Open woodland — tundra; tayga — tundra
210 Warm continental	Mixed forest — coniferous forest — tundra
220 Hot continental	Deciduous or mixed forest — coniferous forest — meadow
230 Subtropical	Mixed forest — meadow
240 Marine	Deciduous or mixed forest — coniferous forest — meadow
250 Prairie	Forest-steppe — coniferous forest — meadow
260 Mediterranean	Mediterranean woodland or shrub — mixed or coniferous forest — steppe or meadow
310 Tropical/subtropical steppe	Steppe or semi-desert — mixed or coniferous forest — alpine meadow or steppe
320 Tropical/subtropical desert	Semi-desert — shrub — open woodland — steppe or alpine meadow
330 Temperate steppe	Steppe — coniferous forest — tundra; steppe — mixed forest — meadow
340 Temperate desert	Semi-desert woodland — meadow
410 Savanna	Open woodland — deciduous forest — coniferous forest — steppe or meadow
420 Rainforest	Evergreen forest — meadow or paramos

Figure 8.14. Montane forest in the Kenya uplands. The tree limb is completely covered with epiphytes. Neg. no. 211362 (Photo by Carl Akeley). Courtesy Department of Library Services, American Museum of Natural History.

ters migrate to the lowest slopes. Year-round residents, such as bears and chipmunks, hibernate.

In mid latitudes, the climates of these montane coniferous forests range from hot and relatively dry to cold, wet, and snowy. The nature of the forest vegetation reflects this wide climatic span. In the United States, the sparse woodland of pinyon and juniper on the desert mountains of southern California and Nevada contrasts strongly with the mossy and cool spruce–fir forest of the cloud-shrouded Great Smoky Mountains of North Carolina and Tennessee.

The subalpine zone is a transitional area between the lush montane forest below and the harsh alpine zone above. It is characterized by scattered, stunted, and misshapen coniferous trees, such as pines, spruces, and hemlocks. As the upper limit of forest, or timberline, is approached, the trees grow in prostrate thickets, called **krummholz**.

Figure 8.15. Open montane forest of ponderosa pine in Coconino National Forest, Arizona. Photograph by Robert G. Bailey.

Animals in this zone are a mixture of the animals found in the alpine zone above and the montane below. Many montane animals move to the higher elevations in summer and retreat down to more protected areas during the winter. The ibex, a goat of the European Alps, lives in the alpine area in summer and the subalpine zone in winter. Year-round residents include the yellow-bellied marmot and the alpine chipmunk.

If the mountain is high enough, the climate of the crest may be too severe for forests. In such places there will be a more or less marked timberline, above which is the alpine zone. The alpine zone has many extreme climatic conditions found in arctic climates, and because the atmosphere is thinner at high altitude, the light intensity is much greater. Most alpine plants grow near the ground where they are protected from the wind. Very few animals inhabit the alpine zone year-round. Those that do, such as the pika, tend to be small because of the scarcity of food. Large mammals, such as the mountain goat and the guanaco, rely on well-insulated coats for protection.

Above the tree line and the alpine zone is the climatic snow line, the boundary of the area covered by rock, and permanent snow and ice (Figure 8.16).

Other climatic elements reinforce the effects of temperature on vertical differentiation. Rainfall, for instance, develops a somewhat vertical zoning. Up to 2 or 3 km of altitude in the middle latitudes, for example, the rainfall on mountain slopes increases. Unbroken mountain ranges are effective barriers to the passage of moisture (Figure 8.17). The mountain ranges along the Pacific Coast of the United States, for example, intercept moisture transported from the Pacific Ocean by prevailing westerly winds, so that coastal ranges are moist and inland regions are dry. However, the effect of mountains in producing rain depends on the character of the air blowing against them. Cold air can

Figure 8.16. Snowfields above valley glaciers near Mount McKinley, Alaska. Photograph by Norman Herkenham, National Park Service.

Figure 8.17.
Annual precipitation correlates well with the altitude of the land in an east–west section at about 38° N in the western United States. From Bailey (1941), p. 192.

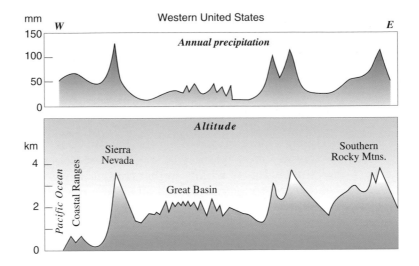

rise only sluggishly and may produce only low stratus (sheet-like) clouds, as along the Peruvian coast (see Figure 7.7, p. 93). The heaviest rains in the world, however, are received on mountain slopes that lie in the path of warm, buoyant, moisture-laden air, as the island of Kauai and the southern slopes of the Himalayas.

East–west mountain ranges act as temperature divides. The lowlands on the poleward side of a mountain range are made colder than they would otherwise be, and the lowlands on the equatorward side are made warmer.

Mountain climates in mid latitudes are an important influence on river flow and floods. The higher ranges serve as snow storage areas, keeping back the precipitation until early or midsummer, releasing it slowly through melting. This maintains continuous river flow.

The upper tree line in mountainous regions is caused by the lack of adequate warmth. The lower treeline found in arid regions is related to the lack of adequate moisture. This combination restricts the growth of forests in arid regions to more or less wide bands along the slopes of mountains (Figure 8.18).

Soils also change their character with increasing altitude, responding to the changes in climate and vegetation. Figure 8.19 shows how soil profiles change with life zones in the western United States.

Man has caused many changes in the mountain ecosystems by mining, agriculture, grazing, and fire. The altitudinal limits of the various forms of human settlement are similar to the horizontal limits of the lowlands. The highest settlements are associated with mining. The next highest settlements are commonly supported by the pasturage of domestic animals. Lower down, the various types of agricultural settle-

Figure 8.18. Altitudinal zonation in the Ruby Mountains in Nevada, with sagebrush semi-desert in the foreground, coniferous forest on the lower mountain slopes, and alpine tundra toward the top. Photograph by Robert G. Bailey.

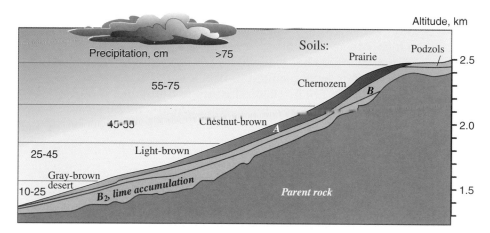

Figure 8.19. Gradation of soils from a dry steppe-climate basin (left) to a cool, humid climate (right) ascending the west slopes of the Big Horn Mountains, Wyoming. (Note that the soil profile is an extreme exaggeration for the purpose of illustration.) From J. Thorp, The effects of vegetation and climate upon soil profiles in northern and northeastern Wyoming, *Soil Science* 32, p. 290, 1931.

ment appear. Both the animals and the crops supporting these settlements show fairly distinct upper limits in any one region. Because sheep can exist on much scantier pasturage than cattle, they are driven highest in the mountains. Cattle and horses usually occupy the richer pastures lower down. Of the crops, the potato reaches the highest altitude. Lower down, the various grains arrange themselves in the expected sequence: barley, rye, wheat, maize, and rice, in descending order. The tropical crops occupy the lower slopes of low-latitude mountains.

In many areas where agriculture has been the main objective, it has been a disaster for the ecosystem. The hill lands of southern China were stripped of their once-rich soils by careless land use. The same has happened in southern Europe, in the Andes, and in the Appalachians and Ozarks of the United States. There are areas, however, where steep slopes have been tilled, using terraces to reduce erosion, for long periods of time without substantial loss. The best examples are found in the Himalayas, the Philippines, and the Andes.

In middle latitudes, with marked seasonal change and fewer vertical zones, mountain people tend to establish their homes at the lower altitude and to ascend each summer with their domestic animals. The movement up and down the mountains in response to seasonal rhythm is called **transhumance**. It occurs on nearly all the lower middle-latitude mountains of the world, except in Japan.

The incidence of forest fires has increased, despite forest-fire protection services restricting the once wide-ranging ground fires to relatively small areas. Exclusion of fires may cause changes in the composition and density of the vegetation, sometimes with disastrous consequences when fuels build up. Many subalpine areas in the Rocky Mountains are now occupied by successional lodgepole pine forest, which covers old burned areas in the spruce–fir zone. Our suppression of wildfires has extended the intervals between major fire events. These efforts have resulted in fires such as the infamous Yellowstone National Park fire in 1988. This fire not only had a different character from past natural fires, but it also burned a far larger area.

Pattern within Zones

The zones give only a broad-brush picture. Variations within a zone break up and differentiate the major, subcontinental zones. For example, the vegetation of the savanna is highly differentiated relative to variation in length of the dry season (Figure 8.20). The geographic patterns of ecosystems within zones caused by these variations are reviewed here; see the author's *Ecosystem Geography* (Bailey 1996) for details.

Within the same macroclimate, broad-scale landforms (geology and

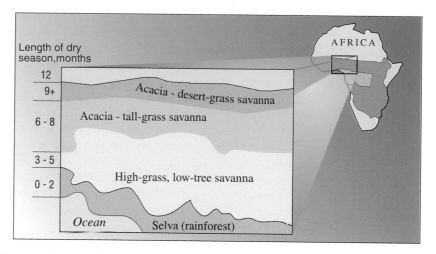

Figure 8.20. Subdivision of the savannas of central Niger. From Shantz and Marbut; from *A Geography of Man*, 2nd ed., by Preston E. James, p. 304. Copyright © 1959 Ginn and Company; reprinted by permission of John Wiley & Sons, Inc.

topography) break up the zonal pattern and provide a basis for further delineation of mesoscale ecosystems, known as **landscapes mosaics**. The same geologic structure in different climates results in different landscapes. For example, limestone in a subarctic climate occurs in depressions and shows intense karstification, while in hot and arid climates, it occurs in marked relief with a few cave tunnels and canyons inherited from colder Pleistocene time (Figure 8.21).

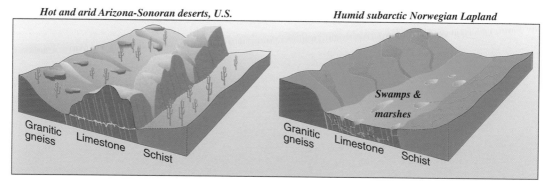

Figure 8.21. Landscape types resulting from similar geology in two different climatic regions. From Corbel (1964), p. 408–409.

A landscape mosaic may be further subdivided into microscale ecosystems called **sites**. Within a landscape, the sites are arranged in a specific pattern. For example, the Idaho Mountains, a temperate-steppe regime highland in the western United States, are made of a complex mosaic of riparian, forest, and grassland sites (Figure 8.22).

Even in areas of uniform macroclimate, topography leads to differences in local climates and soil conditions. Topography causes variations in the amount of solar radiation received, creating **topoclimates** (Thornthwaite 1954), and affects the soil moisture (Figure 8.23).

Variations in drainage, and in steepness of slopes, further affect the soil moisture and biota, in turn creating ecosystem sites. A sequence of moisture regimes, ranging from drier to wetter from the top to the bottom of a slope (Figure 8.24), may be referred to as a soil catena, or a **toposequence** (Major 1951).

Figure 8.22. A mosaic site in the Idaho Mountains. Photograph by John S. Shelton; reproduced with permission.

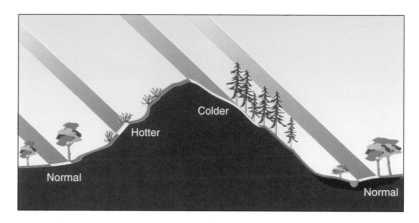

Figure 8.23. Slope and aspect affect temperature, creating topoclimates.

Figure 8.25, in a simplified way, illustrates how topography, even in areas of uniform macroclimate, leads to differences in local climates and soil conditions. The climatic climax theoretically would occur over the entire region but for topography leading to different local climates.

Other topographic, hydrologic, geologic, and/or geochemical deviations may also occur. We can place ecosystem sites into three basic categories: (1) **zonal**, which are typical for the climatic conditions, such as on well-drained sagebrush terraces in a semi-arid climate (Figure 8.26); (2) **azonal**, such as riparian forests; and (3) **intrazonal**, which may occur on extreme types of soil that override the climatic effect, such as very dry sand dunes or black soil over limestone (Figure 8.27).

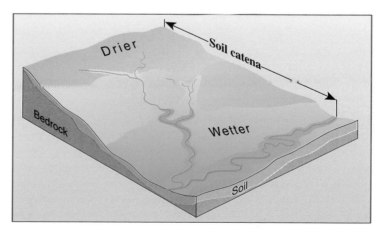

Figure 8.24. Variation is moisture creates a toposequence or catena of soil moisture regimes.

Figure 8.25.
Forest climaxes
relate to topogra-
phy in the temper-
ate continental
zone of southern
Ontario. (Dia-
gram is trun-
cated, showing
only three of nine
possible cli-
maxes.) Simpli-
fied from Hills
(1952).

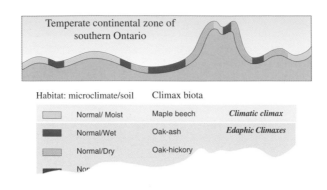

Habitat: microclimate/soil		Climax biota	
	Normal/ Moist	Maple beech	*Climatic climax*
	Normal/Wet	Oak-ash	*Edaphic Climaxes*
	Normal/Dry	Oak-hickory	
	Nor		

Figure 8.26. Zonal sites on sagebrush terraces and azonal riparian forests in Jackson Hole, Wyoming. Photograph by National Park Service.

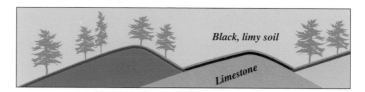

Figure 8.27. An example of an intrazonal site type where a limestone outcrop creates black, limy soil that supports grasses in the midst of a pine forest, Alabama. From Hunt (1974), p. 170.

In summary, the pattern of ecosystems in a region is the product of all these factors, some climatic (resulting from the average state of the atmosphere) and some **edaphic** (resulting from the character of the soil and surface). In general, the broad outlines of the region are the result of climatic conditions, whereas the details observed in a particular place are the result of edaphic conditions.

Figure 8.28. Man-made ecosystem; houses sprout from brown soil in the semiarid steppe near Fort Collins, Colorado. Photograph by Robert G. Bailey.

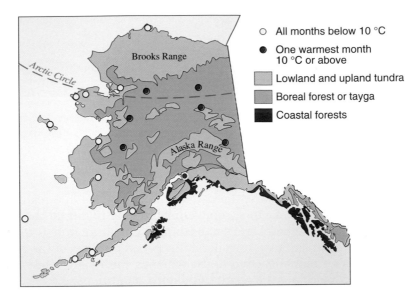

Figure 8.29. The northern and western edges of the boreal forest (tayga) in Alaska correspond closely to a line beyond which all months are below 10°C. Climate data from Walter and Lieth (1960–1967) and Walter et al. (1975); vegetation from Viereck et al. (1992).

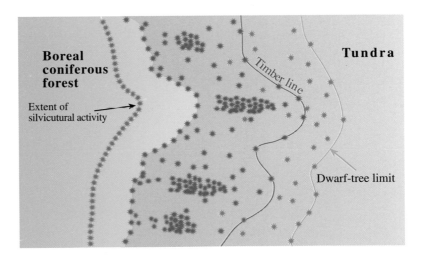

Figure 8.30. The boundary between boreal coniferous forest and tundra is usually a transition zone rather than a sharp line. From Hustich (1953), p. 150, reproduced with permission.

Many smaller natural and man-made ecosystems are incorporated in the larger systems. These systems appear to have their own dynamics, but nevertheless are bound into the surrounding larger systems in many different ways. Three examples scattered through the land ecoregions are rivers, lakes, and towns (Figure 8.28).

Boundaries between Zones

This scheme of defining ecosystems gives a general picture: however, the boundaries may be imprecise. For example, the 10°C isotherm coincides with the northernmost limits of tree growth; hence it separates boreal forest from treeless tundra (Figure 8.29). The observer, of course, would not see an abrupt line but a transition zone—trees on favorable site, muskeg and bog on wetter sites, with tundra on exposed ridges (Figure 8.30).

CHAPTER 9

Summary and Conclusion

In summary, we can interpret the patterns of both ocean and continental ecosystems through the primary factor of climate. Regional ecosystems, or ecoregions, are areas of homogeneous macroclimate.

The arrangement of these ecoregions is regular and predictable *because the controlling factors are the same for each.* We can predict the kind of system that will be found in any particular place on earth if we know the latitude, relative continental or oceanic position, and altitude.

Likewise, within each region is a characteristic pattern of sites that recurs in predictable ways, enabling us to apply the knowledge about one area to another.

This approach of understanding the hierarchy of controls is much more useful than our traditional, scattered, small-scale analyses because all systems operate within a context of larger systems. From our knowledge of the larger systems we can much better understand the smaller systems where we must predict the outcome of land use and natural resource development.

Air Masses and Frontal Zones

A body of air in which the temperature and moisture is fairly uniform over a large area is know as an **air mass**. The boundary between a given air mass and its neighbor is usually sharply defined. This discontinuity is termed a **front**. In the convergence zone between the tropical and polar air masses, winds are variable and high and accompanied by stormy weather. This zone is called the **polar front**. A large number of the Earth's cyclonic storms are generated here. The properties of an air mass are derived partly from the regions over which it passes. Those land or ocean surfaces that strongly impress their characteristics on the overlying air masses are called **source regions**. Air masses are classified according to their latitudinal position (which determines thermal properties) and underlying surface, whether continent or ocean (determining moisture content). They are summarized in Table A.1 and illustrated in Figure A.1.

Table A.1. Properties of air masses[a]

Major group	Subgroup	Source region	Properties of source
Polar (including arctic)	Polar continental (cP)	Arctic Basin; northern Eurasia and northern North America; Antarctica	Cold, dry, very stable
	Polar maritime (mP)	Oceans poleward of 40° or 50°	Cool, moist, unstable
Tropical (including equatorial)	Tropical continental (cT)	Low-latitude deserts, especially Sahara and Australian Deserts	Hot, very dry, stable
	Tropical maritime (mT)	Oceans of tropics and subtropics	Warm, moist, greater instability toward west side of ocean

[a]From Trewartha et al. (1967), p. 105.

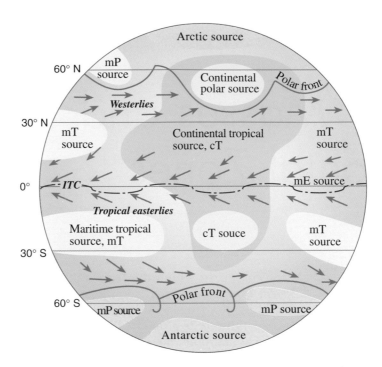

Figure A.1. Source regions of air masses in relation to the polar front and the intertropical convergence zone (ITC). From *Elements of Physical Geography*, 4th ed., by Arthur N. Strahler and Alan H. Strahler, p. 125. Copyright © 1989 by John Wiley & Sons, Inc. Reprinted by permission of John Wiley & Sons, Inc.

Common and Scientific Names

Plants

acacia	*Acacia*
ash	*Fraxinus*
aspen, quaking	*Populus tremuloides*
basswood	*Tilia*
beech, southern	*Nothofagus*
birch	*Betula*
black-eyed Susan	*Rudbeckia hirta*
bluestem, big	*Andropogon gerardii*
bluestem, little	*Schizachyrium scoparium*
buffalograss	*Buchloe dactyloides*
cactus, prickly pear	*Opuntina engelmanni*
cactus, saguaro	*Carnegiea giganteus*
cedro espinoso	*Bombacopsis fendleri*
chestnut, sweet (American)	*Castanea dentata*
creosote bush	*Larrea tridentata*
Douglas fir	*Pseudotsuga menziesii*
elm	*Ulmus*
eucalyptus	*Eucalyptus*
fir	*Abies*
hemlock	*Tsuga*
hemlock, western	*Tsuga heterophylla*
hickory	*Carya*
hornbeam, American	*Carpinus caroliniana*
juniper	*Juniperus*
kauri tree	*Agathis australis*
larch	*Larix*

laurel	*Kalmia*
locoweed	*Oxytropis*
magnolia	*Magnolia*
maple	*Acer*
maple, sugar	*Acer saccharum*
mesquite	*Prosopis*
oak	*Quercus*
oak, cork	*Quercus suber*
ocotillo	*Fouquieria splendens*
pine	*Pinus*
pine, lodgepole	*Pinus contorta*
pine, ponderosa	*Pinus ponderosa*
pinyon	*Pinus edulis*
podocarp tree	*Podocarpus*
popular	*Populus*
red cedar, western	*Thuja plicata*
redwood	*Sequoia sempervirens*
sagebrush	*Artemisia*
smoke tree	*Dalea spinosa*
spruce	*Picea*
sunflower, common	*Helianthus annuus*
tamarisk (salt-cedar)	*Tamarix gallica*
tulip tree	*Liriodendron tulipifera*
walnut	*Juglans*
willow	*Salix*

Animals

antelope	(see pronghorn)
badger, American	*Taxidea taxus*
bear	Ursidae
bear, polar	*Thalarctos maritimus*
beaver	*Castor canadensis*
bison, American	*Bison bison*
bongo	*Boocercus eurycerus*
buffalo, African	*Syncerus caffer*
camel	*Camelus*
caribou	*Rangifer tarandus*
chipmunk	*Eutamias* and *Tamias*
chipmunk, alpine	*Eutamias alpinus*
crocodile	*Crocodylus*
deer	Cervidae
elephant, African	*Loxodonta africana*

elk, American	*Cervus canadensis*
emu	*Dromaius novaehollandiae*
ermine	*Mustela*
fox	*Vulpes* and *Alopex*
gnu (wildebeest)	*Connochaetes taurinus*
goat, North American mountain	*Oreamnos americanus*
guanaco	*Lama guanacoe*
hamster	Cricetidae
hare, arctic	*Lepus arcticus*
hippopotamus	*Hippopotamus amphibius*
ibex	*Capra hircus*
kiwi	*Apteryx*
krill shrimp	*Euphausïa superba*
lemming	*Dicrostonyx* and *Lemmus*
lemur	Lemuridae
lion, mountain (puma, panther)	*Felis concolor*
marmot, yellow-bellied	*Marmota flaviventris*
marten, American	*Martes americana*
mink	*Mustela vison*
moose	*Alces alces*
musk oxen	*Ovibos moschatus*
okapi	*Okapia johnstoni*
panther	(see mountain lion)
penguin	Spheniscidae
pika	*Ochotona princeps*
plover	*Pluvialis* and *Charadrius*
prairie dog	*Cynomys*
pronghorn antelope	*Antilocapra americana*
reindeer	(see caribou)
sabel	(see marten)
sea elephant	*Mirounga leonina*
turkey, wild	*Meleagris gallopavo*
walrus	*Odobenus rosmarus*
whale, blue	*Sibbaldus musculus*
whale, finback	*Balaenoptera physalus*
wolf	*Canis lupus*
zebra	*Equus burchelli*

Conversion Factors

For readers who wish to convert measurements from the metric system of units to the inch–pound–Fahrenheit system, conversion factors are listed below.

Multiply	By	To obtain
millimeters	0.039	inches
centimeters	0.394	inches
meters	3.281	feet
kilometers	0.621	miles
square meters	10.764	square feet
square kilometers	0.386	square miles
hectares	2.471	acres
centigrade	1.8 + 32	Fahrenheit

Glossary of Technical Terms

air mass   a large and essentially homogeneous body of air, many thousands of km² in area, characterized by uniform temperature and humidity

Alfisol   soil order consisting of soils of humid and subhumid climates, with high base status and argillic horizons

alkali   salts found in soils, as in some deserts

anadromous fisheries   migrating from salt water to spawn in fresh water, such as salmon

Aridisol   soil order consisting of soils of dry climates, with or without argillic horizons, and with accumulations of carbonates or soluble salts

arroyo   in Southwest United States, steep-sided dry valley, usually inset in alluvium

azonal   zonal in a neighboring zone but confined to an extra-zonal environment in a given zone; mountains that cut across the lowland ecological zones or regions

biomass   the dry mass of all living material in an area

black prairie soil   (also called *prairie soil* or Brunizem) acid grassland soil

bog   a wet area covered by acid peat

bolson   from a Spanish word, meaning pocket, for a basin of inland drainage

broadleafed   with leaves other than linear in outline; as opposed to needleleafed or grass-like (graminoid) *brown forest soil*   (also called *gray-brown podzolic*) acid soil with dark brown surface layers, rich in humus, grading through lighter colored soil layers to limy parent material; develops under deciduous forest

brown soil   alkaline soil having thin brown surface layer that grades downward into a layer where carbonates have accumulated; develops under grasses and shrubs in semi-arid environments

calcification   accumulation of calcium carbonate in a soil

chaparral   sclerophyll scrub and dwarf forest found throughout the coastal mountain ranges and hills of central and southern California

chernozem   fertile, black or dark brown soil under prairie or grassland with lime layer at some depth between 0.6 and 1.5 m

chestnut-brown soil   short-grass soil in subhumid to semi-arid climate with dark brown layer at top, which is thinner and browner than in chernozem soils, that grades downward to a layer of lime accumulation

climate   generalized statement of the prevailing weather conditions at a given place, based on statistics of a long period of record

climatic climax   the relatively stable community that terminates on zonal soils

climatic regime   seasonality of temperature and moisture

continental shelf   the edge of a continent submerged in relatively shallow seas and oceans

cyclone   whirling storm characteristic of middle latitudes; any rotating low-pressure air system

deciduous   woody plants, or pertaining to woody plants, that seasonally lose all their leaves and become temporarily bare stemmed

delta   the flat alluvial area at the mouth of some rivers where the mainstream splits into several distributaries

desert   supporting vegetation of plants so widely spaced, or sparse, that enough of the substratum shows through to give the dominant tone to the landscape

desertification   degradation of the plant cover and soil as a result of overuse, especially during periods of drought

desert soil   shallow, gray soils containing little humus and excessive amounts of calcium carbonate at depths less than 30 cm

desert-like savanna   tropical semi-desert with scattered low trees or shrubs *division*   as defined for use in this book: a subdivision of a domain determined by isolating areas of definite vegetation affinities that fall within the same regional climate (continents) or areas of similar water temperature, salinity, and currents (oceans)

doldrums   an area near the equator of very ill-defined surface winds associated with the intertropical convergence zone

domain   as defined for use in this book: groups of ecoregions with related climates (continents) or water masses (oceans)

dry savanna or steppe   with six to seven arid months in each year

dry steppe   see *dry savanna*

ecoregion   major ecosystem, resulting from large-scale predictable patterns of solar radiation and moisture, which in turn affect the kinds of local ecosystems and animals and plants found there

ecosystem   an area of any size with an association of physical and biological components so organized that a change in any one component will bring about a change in the other components and in the operation of the whole system

edaphic   resulting from the character of the soil and surface

edaphic climax   stable community of plants that develops on soils different from those supporting a climatic climax

epiphyte   organism that lives on the surface of a plant but does not draw nourishment from it

erg   a very large area of sand dunes within a desert

evergreen   plants, or pertaining to plants, which remain green in parts the year around, either by retaining at least some of their leaves at all times, or by having green stems that carry on photosynthesis

exotic river   stream that flows across a region of dry climate and derives its discharge from adjacent uplands where a water surplus exists

forest   open or closed vegetation with the principal layer consisting of trees averaging more than 5 m in height

forest-steppe   intermingling of steppe and groves or strips of trees

forest-tundra   intermingling of forest and tundra

front   the division between two air masses with different origins and different characteristics

galeria forest   dense tropical, or prairie, forest living along the banks of rivers

geostrophic   pertaining to deflective force due to rotation of the Earth *grassy savanna*   savanna in which woody plants are entirely lacking

gray-brown podzol soil   acid soil under broadleaf deciduous forest; has thin, organic layer over grayish brown, leached layer; layer of deposition is darker brown

hamada   an eroded rock surface found in deserts

Histosol   soil order consisting of soils that are organic

horse latitudes   subtropical high-pressure belt of the oceans

humus   organic material derived, by partial decay, from the organs of dead plants

Inceptisols   soil order consisting of soils with weakly differentiated horizons showing alteration of parent materials

intertropical convergence zone   (commonly abbreviated *ITC*) a broad zone of low pressure, migrating northwards and southwards of the equator with the season, toward which tropical air masses converge

intrazonal   exceptional situations within a zone, e.g., on extreme types of soil that override the climatic effect

isotherm   line on a map connecting points of equal temperature

krummholz   zone of wind-deformed trees between the montane and alpine zones

landscape   see *landscape mosaic*

landscape mosaic   as defined for use in this book: a geographic group of site-level ecosystems

laterite   a residual soil developing in the tropics, containing concentrations of iron and aluminum hydroxides which stain the soil red

laterization   process of forming laterite

latosol   major soil type associated with humid tropics and characterized by red, reddish brown, or yellow coloring

light tayga   tayga forest composed of larch and pine or spruce

macroclimate   large climatic zone arranged in a latitudinal band; climate that lies just above the local modifying irregularities of landform and vegetation

meadow   closed herbaceous vegetation, commonly in stands of rather limited extent, or at least not usually applied to extensive grasslands

meadow steppe   the steppe component of the forest-steppe zone

mixed forest   forest with both needleleafed and broadleafed trees

Mollisol   soil order consisting of soils with a thick, dark-colored, surface-soil horizon, containing substantial amounts of organic matter (humus) and high-base status

monsoon forest   drought-deciduous trees

oceanic polar front   in arctic and antarctic regions, boundary between warm and cold water types, associated with a convergence of surface currents

oceanic whirl   circular movement of air around the subtropical high pressure zone

open woodland   (also called *steppe forest* and *woodland savanna*) open forest with lower layers also open, having the trees or tufts of vegetation discrete, but averaging *less* than their diameter apart

orographic precipitation   precipitation induced by the forced rise of moist air over a mountain barrier

Oxisol   soil order consisting of soils that are mixtures principally of kaolin, hydrates oxides, and quartz

paramo   the alpine belt in the wet tropics

parkland   areas where clumps of trees alternate with grassland, but where neither becomes an extensive, uninterrupted stand

permafrost   permanently frozen soil

piedmont   sequence of landforms along the margins of uplands

plankton   small, floating, or weakly swimming plants and animals, found in salt and fresh water; primarily microscopic algae and protozoa

plant formation class   a world vegetation type dominated throughout by plants of the same life form

playa   a desert lake existing only temporarily after a rain

Pleistocene   the most recent major ice age. Generally the Pleistocene is considered to have begun approximately two million years ago and to have ended eight to ten thousand years ago

podzol   soil order consisting of acid soil in which surface soil is strongly leached of bases and clays

polar front   front lying between cold polar air masses and warm tropical air masses

prairie   consists of tall grasses, mostly exceeding 1 m in height, comprising the dominant herbs, with subdominant forbs (broadleafed herbs)

red-yellow podzol   acid soil under broadleaf deciduous or needleleaf evergreen forest developed in areas of humid subtropical climate

regolith   layer of weathered inorganic and organic debris overlying the surface of the earth

salinity   a saline quality

salinization   precipitation of soluble salts within the soil

savanna   closed grass or other predominantly herbaceous vegetation with scattered or widely spaced woody plants, usually including some low trees

savanna forest   the forest component of the savanna

sclerophyll or sclerophyllous   refers to plants with predominantly hard, stiff leaves that are usually evergreen

selva   an alternative term for tropical rainforest, originally applied to the Amazon Basin

semideciduous forest   composed partly of evergreen and partly deciduous species

semi-desert   (also called *half-desert*) an area of xerophytic shrubby vegetation with a poorly developed herbaceous lower layer, e.g., sagebrush

shrub   a woody plant less than 5 m in height

shrub savanna   closed grass or other predominantly herbaceous vegetation with scattered or widely spaced shrubs

sierozem   see *desert soil*

site   the smallest, or local, ecosystems

small-leafed   as used here, refers to birch and aspen

soil great group   third level of classification of soils, defined by similarities in kind, arrangement, and distinctiveness of horizons, as well as close similarities in moisture and temperature regimes and base status

soil orders   those ten soil classes forming the highest category in the classification of soils

source region   extensive land or ocean surface over which an air mass derives its characteristics

Spodosol   soil order consisting of soils that have accumulations of amorphous materials in subsurface horizons

steppe   (also called *shortgrass prairie*) open herbaceous vegetation, less than 1 m high, with tufts or plants discrete, yet sufficiently close together to dominate the landscape

subtropical high-pressure belts   (also called *cells* or *zones*) belts of persistent high atmospheric pressure tending east-west and centered at about latitude 30° N and S

succession   the replacement of one community of plants and animals by another

tayga   (also spelled *taiga*) a parkland or savanna with needle-leafed (usually evergreen) low trees and shrubs; a Russian word referring to the northern virgin forests

temperate rainforest   dense forest, comprising tall trees, growing in areas of very high rainfall, such as the Pacific Northwest of the United States

thermoisopleth diagram   diagram that shows temperature at a station throughout the day for every day of the year

thermokarst   the formation of a highly irregular ground surface as a result of the thawing of masses of ground ice

topoclimate   the climate of a very small space; influenced by topography

toposequence   a change of a community with topography

trade winds   winds blowing from the east on the equatorward side of the subtropical high-pressure cells

transhumance   the seasonal movement of people and animals to and from fresh pastures

tundra   slow-growing, low formation, mainly closed vegetation of dwarf shrubs, graminoids, and cryptograms, beyond the subpolar or alpine treeline

tundra soil   cold, poorly drained, thin layers of sandy clay and raw humus; without distinctive soil profiles

upwelling   upward motion of cold, nutrient-rich ocean waters, often associated with cool equatorward currents occurring along western continental margins

Ultisol   soil order consisting of soils with horizons of clay accumulation and low base supply

wadi   in Arabia and the Sahara, dry desert valley

westerlies   winds blowing from the west on the poleward side of the subtropical high-pressure cells

woodland   cover of trees whose crowns do not mesh, with the result that branches extend to the ground

xerophyte   a plant adapted to an environment characterized by extreme drought

yellow forest soil   (also called *red-yellow podzol*) soils with weakly developed horizons that also have accumulations of sesquioxides of iron and aluminum; transitional between podzols and latosols

zonal   resulting from the average state of the atmosphere; variation in environmental conditions in a north–south direction

zonal soil   well-developed deep soils on moderate surface slopes that are well drained

Notes

[1]Terms in bold are defined in the Glossary, p. 143.

[2]Great soil group according to the 1938 system of soil classification (U.S. Department of Agriculture 1938). The most nearly equivalent orders of the new soil taxonomy (Soil Survey Staff 1975) are given in parentheses. Described in the Glossary, p. 143.

[3]Zones of latitude may be described as follows: from the equator to 30° are *low latitudes*; from 30° to 60° are the *middle latitudes*; from 60° to the poles are the *high latitudes*.

[4]In this book tayga, often spelled taiga, will be written with a "y" following the Russian.

[5]Other methods for mapping zones at the global scale are those of Thornthwaite (1931, 1933), Holdridge (1947), and Walter and Box (1976). All methods appear to work better in some areas than in others and to have gained their own adherents. The Köppen system was chosen as the basis for ecoregion delineation because it has become the international standard for geographical purposes.

Selected Bibliography

On Ecological Divisions of the Oceans

Bogorov, V.G. 1962. Problems of the zonality of the world ocean. In: Harris, C.D., ed. *Soviet Geography, Accomplishments and Tasks*. Occasional Publ. No. 1. New York: American Geographical Society: 188–194.

Dietrich, G. 1963. *General Oceanography: An introduction*. New York: John Wiley. 588 pp.

Elliott, F.E. 1954. The geographic study of the oceans. In: James, P.E.; Jones, C.F., eds. *American geography: Inventory & prospect*. Syracuse: Association of American Geographers by Syracuse University Press: 410–426.

Hayden, B.P.; Ray, G.C.; Dolan, R. 1984. Classification of coastal and marine environments. *Environmental Management* 11:199–207.

James, P.E. 1936. The geography of the oceans: A review of the work of Gerhard Schott. *Geographical Review* 26:664–669.

Joerg, W.L.G. 1935. The natural regions of the world oceans according to Schott. In: *Transactions of the American Geophysical Union, Sixteenth Annual Meeting, Part I.* April 25–26, 1935; Washington, DC: 239–245.

Schott, G. 1936. Die aufteilung der drei ozeane in natürliche regionen. *Petermann's Mitteilungen* 82:165–170; 218–222.

Sherman, K.; Alexander, L.M.; and Gold, B.D., eds. 1990. *Large Marine Ecosystem: Patterns, Processes, and Yields*. Washington, DC: American Association for the Advancement of Science. 242 pp.

Terrell, T.T. 1979. *Physical Regionalization of Coastal Ecosystems of the United States and Its Territories*. FWS/OBS-78/80. Washington, DC: U.S. Fish and Wildlife Service. 30 pp.

On the Climatic and Ecological Divisions of the Continents as a Whole or Larger Parts

Akin, W.E. 1991. *Global Patterns: Climate, Vegetation, and Soils.* Norman: University of Oklahoma Press. 370 pp.

Allee, W.C; Schmidt, K.P. 1951. *Ecological Animal Geography* (based on *Tiergeographie auf oekologischer Grundlage* by Richard Hesse). 2d ed. New York: John Wiley. 715 pp.

Archibold, O.W. 1995. *Ecology of World Vegetation.* London: Chapman & Hall. 510 pp.

Atwood, W.W. 1940. *The Physiographic Provinces of North America.* Boston: Ginn. 536 pp.

Austin, M.E. 1965. *Land Resource Regions and Major Land Resource Areas of the United States (exclusive of Alaska and Hawaii).* Agric. Handbook 296. Washington, DC: USDA Soil Conservation Service. 82 pp. with separate map at 1:7,500,000.

Bailey, R.G. 1976. *Ecoregions of the United States.* Ogden, UT: USDA Forest Service, Intermountain Region. 1:7,500,000; colored.

Bailey, R.G. 1983. Delineation of ecosystem regions. *Environmental Management* 7:365–373.

Bailey, R.G. 1984. Testing an ecosystem regionalization. *Journal of Environmental Management* 19:239–248.

Bailey, R.G. 1985. The factor of scale in ecosystem mapping. *Environmental Management* 9:271–276.

Bailey, R.G. 1988a. *Ecogeographic Analysis: A Guide to the Ecological Division of Land for Resource Management.* Misc. Publ. No. 1465. Washington, DC: USDA Forest Service. 16 pp.

Bailey, R.G. 1988b. Problems with using overlay mapping for planning and their implications for geographic information systems. *Environmental Management* 12:11–17.

Bailey, R.G. 1989. Explanatory supplement to ecoregions map of the continents. *Environmental Conservation* 16:307–309 with separate map at 1:30,000,000.

Bailey, R.G. 1991. Design of ecological networks for monitoring global change. *Environmental Conservation* 18:173–175.

Bailey, R.G. 1995. *Description of the Ecoregions of the United States.* 2d ed. rev. and expanded (1st ed. 1980). Misc. Publ. No. 1391 (rev.). Washington, DC: USDA Forest Service. 108 pp. with separate map at 1:7,500,000.

Bailey, R.G. 1996. *Ecosystem Geography.* New York: Springer-Verlag. 204 pp.

Bailey, R.G.; Cushwa, C.T. 1981. *Ecoregions of North America.* FWS/OBS-81/29. Washington, DC: U.S. Fish and Wildlife Service. 1:12,000,000; colored.

Barnes, B.V. 1984. Forest ecosystem classification and mapping in Baden-Württemberg, West Germany. In: Bockheim, J.G., ed. *Proceedings, Forest Land Classification: Experiences, Problems, Perspectives.* March 18–20, 1984. Madison, Wisconsin: 49–65.

Bashkin, V.N.; Bailey, R.G. 1993. Revision of map of ecoregions of the world (1992–1995). *Environmental Conservation* 20:75–76.

Bear, F.E.; Pritchard, W.; Akin, W.E. 1986. *Earth: The Stuff of Life.* 2d ed. Norman: University of Oklahoma Press. 318 pp.

Beckinsale, R.P. 1971. River regimes. In: Chorley, R.J., ed. *Introduction to Physical Hydrology.* London: Methuen: 176–192.

Bennett, C.F. 1975. *Man and Earth's Ecosystems.* New York: John Wiley. 331 pp.

Berg, L.S. 1947. *Geograficheskiye zony Sovetskogo Soyuza* (Geographical zones of the Soviet Union). vol. 1, 3rd ed. Geografgiz, Moscow.

Biasutti, R. 1962. *Ll Paesaggio Terrestre.* 2d ed. Torino: Unione Tipografico. 586 pp.

Billings, W.D. 1964. *Plants and the Ecosystem.* Belmont, CA: Wadsworth. 154 pp.

Birot, P. 1970. *Les Régions Naturelles du Globe.* Paris: Masson. 380 pp.

Bourne, R. 1931. *Regional Survey and Its Relation to Stocktaking of the Agricultural and Forest Resources of the British Empire.* Oxford Forestry Memoirs 13, Clarendon Press. 169 pp.

Bowman, I. 1911. *Forest Physiography, Physiography of the U.S. and Principal Soils in Relation to Forestry.* New York: John Wiley. 759 pp.

Brazilevich, N.I.; Rodin, L.Ye.; Rozov, N.N. 1971. Geographical aspects of biological productivity. *Soviet Geography* 12:293–317.

Breymeyer, A.I. 1981. Monitoring of the functioning of ecosystems. *Environmental Monitoring and Assessment* 1:175–183.

Brown, D.E.; Lowe, C.H.; Pase, C.P. 1980. *A Digitized Systematic Classification for Ecosystems with an Illustrated Summary of the Natural Vegetation of North America.* USDA Forest Service General Technical Report RM-73. Fort Collins, CO: Rocky Mountain Forest and Range Experiment Station. 93 pp.

Budyko, M.I. 1974. *Climate and Life* (English edition by D.H. Miller). New York: Academic Press. 508 pp.

Burger, D. 1976. The concept of ecosystem region in forest site classification. In: *Proceedings, International Union of Forest Research Organizations (IUFRO).* XVI World Congress, Division I; 20 June–2 July 1976; Oslo, Norway. Oslo, Norway: IUFRO: 213–218.

Christopherson, R.W. 1994. *Geosystems: An Introduction to Physical Geography.* 2d ed. Englewood Cliffs, NJ: Macmillan. 663 pp.

Cleland, D.T; Avers, P.E.; McNab, W.H.; Jensen, M.E.; Bailey, R.G.; King, T.; Russell, W.E. 1997. National hierarchical framework of ecological units. In: Boyce, M.S.; Haney, A., eds. *Ecosystem Management.* New Haven: Yale University Press: 181–200.

Corbel, J. 1964. L'érosion terrestre étude quantitative (méthodes-techniques-résultats) *Annales de Geographie* 73:385–412.

Cowardin, L.M.; Carter, V.; Golet, F.C.; LaRoe, E.T. 1979. *Classification of Wetlands and Deep-Water Habitats of the United States.* FWS/OBS-79/31. Washington, DC: U.S. Fish and Wildlife Service. 103 pp.

Crowley, J.M. 1967. Biogeography. *Canadian Geographer* 11:312–326.

Dansereau, P. 1957. *Biogeography—An Ecological Perspective.* New York: Ronald Press. 394 pp.

de Laubenfels, D.J. 1970. *A Geography of Plants and Animals.* Dubuque, IA: Wm. C. Brown. 133 pp.

de Laubenfels, D.J. 1975. *Mapping the World's Vegetation: Regionalization of Formations and Flora.* Syracuse: Syracuse University Press. 246 pp.

Delvaux, J.; Galoux, A. 1962. *Les Territoires Écologiques du Sud-Est Belge.* Centre d'Ecologie generale. Travaux hors śerie. 311 pp.

Dokuchaev, V.V. 1899. *On the Theory of Natural Zones. Sochineniya* (Collected Works). v. 6. Moscow-Leningrad: Academy of Sciences of the USSR, 1951.

Ecoregion Working Group. 1989. *Ecoclimatic Regions of Canada, First Approximation.* (Ecological Land Classif. Series No. 23.) Ottawa: Environment Canada. 119 pp. with separate map at 1:7,500,000.

Eyre, S.R. 1963. *Vegetation and Soils: A World Picture.* Chicago: Aldine Publishing. 324 pp.

FAO/UNESCO. 1971–1978. *FAO/UNESCO Soil Map of the World 1:5 Million. North America, South America, Mexico and Central America, Europe, Africa, South Asia, North and Central Asia, Australia.* Paris: UNESCO.

Fenneman, N.M. 1928. Physiographic divisions of the United States. *Annals, Association of American Geographers* 18:261–353.

Forman, R.T.T. 1995. *Land Mosaics: The Ecology of Landscapes and Regions.* Cambridge: Cambridge University Press. 632 pp.

Forman, R.T.T.; Godron, M. 1986. *Landscape Ecology.* New York: John Wiley. 619 pp.

Garner, H.F. 1974. *The Origins of Landscapes: A Synthesis of Geomorphology.* New York: Oxford University Press. 734 pp.

Gaussen, H. 1954. *Théorie et Classification des Climats et Microclimats.* 8me Congr. Internat. Bot. Paris, Sect. 7 et 3, pp. 125–130.

Geiger, R. 1965. *The Climate Near the Ground* (trans.). Cambridge, MA: Harvard University Press. 611 pp.

Gerasimov, I.P., ed. 1964. Types of natural landscapes of the Earth's land areas. Plate 75. In: *Fiziko-Geograficheskii Atlas Mira* (Physico-geographic atlas of the world). Moscow: USSR Acad. Sci. and Main Administration of Geodesy and Cartography. 1:80,000,000.

Gersmehl, P.; Napton, D.; Luther, J. 1982. The spatial transferability of resource interpretations. In: Braun, T.B., ed. *Proceedings, National In-Place Resource Inventories Workshop.* August 9–14, 1981. University of Maine, Orono. Washington, DC: Society of American Foresters: 402–405.

Gleason, H.A.; Cronquist, A. 1964. *Natural Geography of Plants.* New York: Columbia University Press. 420 pp.

Goudie, A. 1993. *The Nature of the Environment.* 3rd ed. Oxford: Blackwell. 397 pp.

Grigor'yev, A.A. 1961. The heat and moisture regime and geographic zonality. *Soviet Geography: Review and Translation* 2:3–16.

Günther, M. 1955. Untersuchungen über das Ertragsvermögen der Hauptholzarten in Bereich verschiederner des württenbergischen Necharlandes. *Mitt. Vereins f. forstl. Standortsk. u. Forstpflz.* 4:5–31.

Haggett, P. 1972. *Geography: A Modern Synthesis.* New York: Harper & Row. 483 pp.

Hammond, E.H. 1954. Small-scale continental landform maps. *Annals Association of American Geographers* 44:33–42.

Hare, T. 1994. *Habitats: 14 Gatefold Panoramas of the World's Ecological Zones.* New York: Macmillan. 143 pp.

Herbertson, A.J. 1905. The major natural regions: an essay in systematic geography. *Geography Journal* 25:300–312.

Hidore, J.J. 1974. *Physical Geography: Earth Systems.* Glenview, IL: Scott, Foresman and Co. 418 pp.

Hills, A. 1952. *The Classification and Evaluation of Site for Forestry.* Res. Rep. 24. Toronto: Ontario Department of Lands and Forest. 41 pp.

Hills, G.A. 1960. Comparison of forest ecosystems (vegetation and soil) in different climatic zones. *Silva Fennica* 105:33–39.

Holdridge, L.R. 1947. Determination of world plant formations from simple climatic data. *Science* 105:367–368.

Hole, F.D. 1978. An approach to landscape analysis with emphasis on soils. *Geoderma* 21:1–23.

Hole, F.D.; Campbell, J.B. 1985. *Soil Landscape Analysis.* Totowa, NJ: Rowman & Allanheld. 196 pp.

Hopkins, A.D. 1938. *Bioclimatics: A Science of Life and Climate Relations.* Misc. Publ. No. 280. Washington, DC: U.S. Department of Agriculture. 188 pp.

Host, G.E.; Polzer, P.L.; Mladenoff, D.J.; White, M.A.; Crow, T.R. 1996. A quantitative approach to developing regional ecosystem classifications. *Ecological Applications* 6:608–618.

Howard, J.A.; Mitchell, C.W. 1985. *Phytogeomorphology.* New York: John Wiley. 222 pp.

Huggett, R.J. 1995. *Geoecology: An Evolutionary Approach.* London: Routledge. 320 pp.

Hunt, C.B. 1974. *Natural Regions of the United States and Canada.* San Francisco: W.H. Freeman. 725 pp.

Isachenko, A.G. 1973. *Principles of Landscape Science and Physical-Geographical Regionalization* (trans. from Russian by R.J. Zatorski, edited by J.S. Massey). Carlton, Victoria, Australia: Melbourne University Press. 311 pp.

James, P.E. 1959. *A Geography of Man.* 2d ed. Boston: Ginn. 656 pp.

Jenny, H. 1941. *Factors of Soil Formation.* New York: McGraw-Hill. 281 pp.

Joerg, W.L.G. 1914. The subdivision of North America into natural regions: a preliminary inquiry. *Annals Association of American Geographers* 4:55–83.

King, L.C. 1967. *The Morphology of the Earth: A Study and Synthesis of World Scenery.* 2d ed. New York: Hafner. 726 pp.

Klijn, F.; Udo de Haes, H.A. 1994. A hierarchical approach to ecosystems and its applications for ecological land classification. *Landscape Ecology* 9:89–104.

Köppen, W. 1931. *Grundiss der Klimakunde.* Berlin: Walter de Gruyter. 388 pp.

Krajina, V.J. 1965. Biogeoclimatic zones and classification of British Columbia. In: Krajina, V.J., ed. *Ecology of Western North America.* Vancouver, British Columbia: University of British Columbia Press: 1–17.

Küchler, A.W. 1964. *Potential Natural Vegetation of the Conterminous United States*. Spec. Pub. 36. New York: American Geographical Society. 116 pp. with separate map at 1:3,168,000.

Küchler, A.W. 1973. Problems in classifying and mapping vegetation for ecological regionalization. *Ecology* 54:512–523.

Küchler, A.W. 1974. Boundaries on vegetation maps. In: Tüxen, R., ed. *Tatsachen und Probleme der Grenzen in der Vegetation*. Lehre, Germany: Verlag von J. Cramer: 415–427.

Kul'batskaya, I.Y., editor-in-chief. 1988. *Geograficheskiye Poyasa i Zonal'nyye Tipy Landshaftov Mira* (Geographic belts and zonal types of landscapes of the world). USSR Academy of Sciences and Main Administration of Geodesy and Cartography. Moscow, USSR (in Russian). 1:15,000,000.

Leser, H. 1976. *Landscaftsökologie*. Stuttgart: Eugen Ulmer. 432 pp.

Lewis, G.M. 1966. Regional ideas and reality in the Cis-Rocky Mountain west. *Transactions of the Institute of British Geographers* 38:135–150.

Lieth, H. 1964–1965. *A Map of Plant Productivity of the World*. Wiesbaden: Geographisches Taschenbuch: 72–80.

MacArthur, R.H. 1972. *Geographical Ecology: Patterns in the Distribution of Species*. Princeton, NJ: Princeton University Press. 269 pp.

Major, J. 1951. A functional, factorial approach to plant ecology. *Ecology* 32:392–412.

Mather, J.R.; Sdasyuk, G.V., eds. 1991. *Global Change: Geographical Approaches*. Tucson: University of Arizona Press. 289 pp.

McHarg, I.L. 1969. *Design with Nature*. Garden City, NY: American Museum of Natural History by The Natural History Press. 197 pp.

Merriam, C.H. 1898. *Life Zones and Crop Zones of the United States*. (Bull. Div. Biol. Surv. 10.) Washington, DC: U.S. Department of Agriculture: 1–79.

Milanova, E.V.; Kushlin, A.V., eds. 1993. *World Map of Present-Day Landscapes: An Explanatory Note*. Moscow: Moscow State University. 33 pp. with separate map at 1:15,000,000.

Mil'kov, F.N. 1979. The contrastivity principle in landscape geography. *Soviet Geography* 20:31–40.

Miller, D.H. 1978. The factor of scale: ecosystem, landscape mosaic, and region. In: Hammond, K.A.; Macinko, G.; and Fairchild, W.B., eds. *Sourcebook on the Environment*. Chicago: University of Chicago Press: 63–88.

Mitchell, C.W. 1973. *Terrain Evaluation*. London: Longman. 221 pp.

Müller, P. 1974. *Aspects of Zoogeography*. The Hague: Dr. W. Junk. 208 pp.

Muller, R.A.; Oberlander, T.M. 1978. *Physical Geography Today: A Portrait of a Planet*. 2d ed. New York: Random House. 590 pp.

Nielson, R.P. 1987. Biotic regionalization and climatic controls in western North America. *Vegetatio* 70:135–147.

Noss, R.F. 1983. A regional landscape approach to maintaining diversity. *BioScience* 33:700–706.

Odum, E.P. 1971. *Fundamentals of Ecology*. 3rd ed. Philadelphia: W.B. Saunders. 574 pp.

Olson, J.S.; Watts, J.S. 1982. Major world ecosystem complexes. In: *Carbon in Live Vegetation of Major World Ecosystems*. ORNL-5862. Oak Ridge, TN: Oak Ridge National Laboratory. 1:30,000,000.

Oosting, H.J. 1956. *The Study of Plant Communities*. 2d ed. San Francisco: W.H. Freeman. 440 pp.

Passarge, S. 1929. *Die Landschaftsgürtel der Erde, Natur und Kultur*. Breslau: Ferdinand Hirt. 144 pp.

Pflieger, W.L. 1971. *A Distributional Study of Missouri Fishes*. Univ. Kansas Publ., Mus. Nat. Hist. 20:225–570.

Pojar, J.; Klinka, K.; Meidinger, D.V. 1987. Biogeoclimatic ecosystem classification in British Columbia. *Forest Ecology and Management* 22:119–154.

Rowe, J.S. 1962. The geographic ecosystem. *Ecology* 43:575–576.

Rowe, J.S.; Sheard, J.W. 1981. Ecological land classification: a survey approach. *Environmental Management* 5:451–464.

Sauer, C.O. 1925. The morphology of landscape. *University of California Publications in Geography* 2:19–53.

Schultz, J. 1995. *The Ecozones of the World: The Ecological Divisions of the Geosphere* (trans. from German by I. and D. Jordan). Berlin: Springer-Verlag. 449 pp.

Schmithüsen, J. 1976. *Atlas zur Biogeographie*. Mannheim-Wien-Zurich: Bibliographisches Institut. 88 pp.

Shantz, H.L.; Marbut, C.F. 1923. *The Vegetation and Soils of Africa*. Research Series 13. New York: American Geographical Society. 263 pp.

Shelford, V.E. 1963. *The Ecology of North America*. Urbana: University of Illinois Press. 609 pp.

Smith, R.L. 1996. *Ecology and Field Biology*. 5th ed. New York: HarperCollins. 733 pp.

Soil Survey Staff. 1975. *Soil Taxonomy: A Basic System for Making and Interpreting Soil Surveys*. Agric. Handb. 436. Washington, DC: U.S. Department of Agriculture. 754 pp.

Stephenson, N.L. 1990. Climatic control of vegetation distribution: the role of the water balance. *American Naturalist* 135:649–670.

Strahler, A.N. 1965. *Introduction to Physical Geography*. New York: John Wiley. 455 pp.

Strahler, A.N.; Strahler, A.H. 1989. *Elements of Physical Geography*. 4th ed. New York: John Wiley. 562 pp.

Sukachev, V.; Dylis, N. 1964. *Fundamentals of Forest Biogeocoenology* (trans. from Russian by J.M. Maclennan). London: Oliver & Bond. 672 pp.

Swanson, F.J.; Kratz, T.K.; Caine, N.; Woodmansee, R.G. 1988. Landform effects on ecosystem patterns and processes. *Bioscience* 38:92–98.

Thornthwaite, C.W. 1931. The climates of North America according to a new classification. *Geographical Review* 21:633–655 with separate map at 1:20,000,000.

Thornthwaite, C.W. 1933. The climates of the Earth. *Geographical Review* 23:433–440 with separate map at 1:77,000,000.

Thornthwaite, C.W. 1941. Climate and settlement in the Great Plains. In: *Climate and Man, 1941 Yearbook of Agriculture*. Washington, DC: U.S. Government Printing Office: 177–196.

Thornthwaite, C.W. 1954. Topoclimatology. In: *Proceedings of the Toronto Meteorological Conference, September 9–15, 1953*. Toronto: Royal Meterological Society: 227–232.

Trewartha, G.T. 1968. *An Introduction to Climate*. 4th ed. New York; McGraw-Hill. 408 pp.

Trewartha, G.T.; Robinson, A.H.; Hammond, E.H. 1967. *Physical Elements of Geography*. 5th ed. New York: McGraw-Hill. 527 pp.

Tricart, J.; Cailleux, A. 1972. *Introduction to Climatic Geomorphology* (trans. from French by C.J. Kiewiet de Jonge). New York: St. Martin's Press. 274 pp.

Troll, C. 1964. Karte der jahrzeiten-klimate der erde. *Erdkunde* 17:5–28.

Troll, C. 1966. Seasonal climates of the earth. The seasonal course of natural phenomena in the different climatic zones of the earth. In: Rodenwaldt, E.; Jusatz, H.J., eds. *World Maps of Climatology*, 3rd ed. Berlin: Springer-Verlag: 15–28 with separate map at 1:45,000,000 by C. Troll and K.H. Paffen.

Troll, C. 1971. Landscape ecology (geoecology) and biogeocenology—a terminology study. *Geoforum* 8:43–46.

Tukhanen, S. 1986. Delimitation of climatic-phytogeographical regions at the high-latitude area. *Nordia* 20:105–112.

Udvardy, M.D.F. 1975. *A Classification of the Biogeographical Provinces of the World*. Occasional Paper No. 18. Morges, Switzerland: International Union for Conservation of Nature and Natural Resources. 48 pp.

U.S. Department of Agriculture. 1938. *Soils and Men, 1938 Yearbook of Agriculture*. Washington, DC: U.S. Government Printing Office. 1232 pp.

Vale, T.R. 1982. *Plants and People: Vegetation Change in North America*. Washington, DC: Association of American Geographers. 88 pp.

Veatch, J.O. 1930. Natural geographic divisions of land. *Michigan Academy of Sciences, Arts and Letters*. 19:417–427.

Volubief, V.P. 1953. *Soils and Climate*. Baku: Azerbaijan Academy of Science. 319 pp.

Walter, H. 1977. *Vegetationszonen und Klima: Die Ökologische Gliederung der Biogeosphäre*. Stuttgart, Germany: Eugen Ulmer Verlag. 309 pp.

Walter, H. 1984. *Vegetation of the Earth and the Ecological Systems of the Geo-Biosphere* (trans. from German by O. Muise). 3rd ed. Berlin: Springer-Verlag. 318 pp.

Walter, H.; Box. E. 1976. Global classification of natural terrestrial ecosystems. *Vegetatio* 32:75–81.

Walter, H; Harnickell, E.; Mueller-Dombois, D. 1975. *Climate-Diagram Maps of the Individual Continents and the Ecological Climatic Regions of the Earth*. Berlin: Springer-Verlag. 36 pp. with 9 maps.

Walter, H.; Breckle, S.W. 1985. *Ecological Systems of the Geobiosphere. Vol. 1. Ecological Principles in Global Perspective* (trans. from German by S. Gruber). Berlin: Springer-Verlag. 242 pp.

Walter, H.; Lieth, H. 1960–1967. *Klimadiagramm Weltatlas*. Jena, East Germany: G. Fischer Verlag. maps, diagrams, profiles. Irregular pagination.

Whittaker, R.H. 1975. *Communities and Ecosystems*. 2d ed. New York: MacMillan. 387 pp.

World ecoregions, types of natural landscapes. 1995. In: Espenshade, E.B.; Hudson, J.C; Morrison, J.L., eds. *Goode's World Atlas*. 19th ed. Chicago: Rand McNally: 22–23. 1:77,000,000.

Yoshino, M.M. 1975. *Climate in a Small Area: An Introduction to Local Meteorology*. Tokyo: University of Tokyo Press. 549 pp.

On Regional Systems

Polar

Flint, R.F. 1971. *Glacial and Quaternary Geology*. New York: John Wiley. 892 pp.

Hare, F.K. 1950. Climate and zonal divisions of the boreal forest formation in eastern Canada. *Geographical Review* 40:615–635.

Hare, F.K.; Ritchie, J.C. 1972. The boreal bioclimates. *Geographical Review* 62:333–365.

Hustich, I. 1953. The boreal limits of conifers. *Arctic* 6:149–162.

Ives, J.D.; Barry, R.G., eds. 1974. *Arctic and Alpine Environments*. London: Methuen. 999 pp.

Pielke, R.A.; Vidale, P.L. 1995. The boreal forest and the polar front. *Journal of Geophysical Research* 100:25755–25758.

Polunin, N. 1951. The real Arctic: Suggestions for its delimitation, subdivision and characterization. *Journal of Ecology* 39:308–315.

Schultz, J. 1995. see Chapters 3.1 and 3.2.

Shear, J.A. 1964. The polar marine climate. *Annals Association of American Geographers* 54:310–317.

Tricart, J. 1970. *Geomorphology of Cold Environments* (trans. from French by Edward Watson). London: Macmillan. 320 pp.

Viereck, L.A.; Dyrness, C.T.; Batten, A.R.; Wenzlick, K.L. 1992. *The Alaska Vegetation Classification*. General Technical Report PNW-GTR-286. Portland, OR: USDA Forest Service, Pacific Northwest Research Station. 278 pp.

Humid Temperate

Albert, D.A.; Denton, S.R.; Barnes, B.V. 1986. *Regional Landscape Ecosystems of Michigan*. Ann Arbor: University of Michigan. 32 pp. with separate map at 1:1,000,000.

Borchert, J.F. 1950. The climate of the central North American grassland. *Annals Association of American Geographers* 40:1–39.

Di Castri, F.; Goodall, D.W.; Specht, R.L., eds. 1981. *Mediterranean-Type Shrublands*. Ecosystems of the World 11. Amsterdam: Elsevier. 643 pp.

Dix, R.L; Smeins, F.E. 1967. The prairie, meadow, and marsh vegetation of Nelson County, North Dakota. *Canadian Journal of Botany* 45:21–58.

Orme, A.T.; Bailey, R.G. 1971. Vegetation and channel geometry in Monroe Canyon, southern California. *Yearbook of the Association of Pacific Coast Geographers* 33:65–82.

Ovington, J.D., ed. 1983. *Temperate Broad-Leaved Evergreen Forests*. Ecosystems of the World 10. Amsterdam: Elsevier. 241 pp.

Röhrig, E.; Ulrich, B., eds. *Temperate Deciduous Forests*. Ecosystems of the World 7. Amsterdam: Elsevier. 635 pp.

Schultz, J. 1995. see Chapters 3.3, 3.6, and 3.8.

Dry

Bailey, R.W. 1941. Climate and settlement of the arid region. In: *1941 Year-book of Agriculture*. Washington, DC: U.S. Department of Agriculture: 188–196.

Coupland, R.T., ed. 1992, 1993. *Natural Grasslands*. Ecosystems of the World 8A and 8B. Amsterdam: Elsevier. 469 pp. and 556 pp.

Evenari, M.; Noy-meir, I.; Goodall, D.W., eds. 1985, 1986. *Hot Deserts and Arid Shrublands*. Ecosystems of the World 12A and B. Amsterdam: Elsevier. 365 pp. and 451 pp.

Goudie, A.S.; Wilkinson, J.C. 1977. *The Warm Desert Environment*. Cambridge: Cambridge University Press. 88 pp.

Hunt, C.B. 1966. *Plant Ecology of Death Valley, California*. Professional Paper 509. Washington, DC: U.S. Geological Survey. 68 pp.

Schultz, J. 1995. see Chapters 3.4 and 3.5.

Shreve, F. 1942. The desert vegetation of North America. *Botanical Review* 8:195–246.

UNESCO. 1977. *Map of the World Distribution of Arid Regions*. MAB Technical Notes 7. Paris: United Nations Education, Scientific and Cultural Organization. 54 pp. with separate map at 1:25,000,000.

West, N.E., ed. 1983. *Temperate Deserts and Semi-Deserts*. Ecosystems of the world 5. Amsterdam: Elsevier. 522 pp.

Humid Tropical

Beard, J.S. 1955. The classification of tropical American vegetation types. *Ecology* 36:89–100.

Bourliere, F., ed. 1983. *Tropical Savannas*. Ecosystems of the world 13. Amsterdam: Elsevier. 730 pp.

Cole, M.M. 1960. Cerrado, caatinga and pantanal: the distribution and origin of the savanna vegetation of Brazil. *Geographical Journal* 126:168–179.

Fosberg, F.R.; Garnier, B.J.; Küchler, A.W. 1961. Delimitation of the humid tropics. *Geographical Review* 51:333–347 with separate map at 1:60,000,000.

Golley, F.B., ed. 1983. *Tropical Rain Forest Ecosystems*. Ecosystems of the world 14A. Amsterdam: Elsevier. 381 pp.

le Houerou, H.N.; Popov, G.F. 1981. *An Eco-Climatic Classification of Intertropical Africa*. FAO Technical Paper 31. Rome: FAO. 40 pp.

Ruhe, R.V. 1960. Elements of the soil landscape. *7th International Congress of Soil Science* 23:165–170.

Schultz, J. 1995. see Chapters 3.7 and 3.9.

Tosi, J.S. 1964. Climatic control of terrestrial ecosystems: a report on the Holdridge model. *Economic Geography* 40:173–181.

Tricart, J. 1972. *The Landforms of the Humid Tropics, Forests, and Savannas* (trans. from French by Conrad J. Kiewiet de Jonge). London: Longman. 306 pp.

Mountains

Barry, R.G. 1992. *Mountain Weather and Climate.* 2d ed. London: Routledge. 402 pp.

Daubenmire, R. 1943. Vegetation zonation in the Rocky Mountains. *Botanical Review* 9:325–393.

Merriam, C.H. 1890. Results of a biological survey of the San Francisco Mountain region and desert of the Little Colorado, Arizona. *North American Fauna* 3:1–136.

Peet, R.K. 1981. Forest vegetation of the Colorado Front Range. *Vegetatio* 45:3–75.

Pfister, R.D.; Arno, S.F. 1980. Classifying forest habitat types based on potential climax vegetation. *Forest Science* 26:52–70.

Swan, L. W. 1967. Alpine and aeolian regions of the world. In: Wright, Jr., H.E.; Osburn, W.H., eds. *Arctic and Alpine Environments.* Bloomington, IN: Indiana University Press: 29–54.

Thorp, J. 1931. The effects of vegetation and climate upon soil profiles in northern and northeastern Wyoming. *Soil Science* 32:283–302.

Troll, C. 1968. The Cordilleras of the tropical Americas, aspects of climatic, phytogeographical and agrarian ecology. In: Troll, C., ed. *Geo-Ecology of the Mountainous Regions of the Tropical Americas.* Bonn: Ferd. Dümmlers Verlag: 15–56.

Troll, C. 1972. Geoecology and the world-wide differentiation of high-mountain ecosystems. In: Troll, C., ed. *Geoecology of the High-Mountain Regions of Eurasia.* Wiesbaden: Franz Steiner Verlag: 1–21.

About the Author

ROBERT G. BAILEY (b. 1939) is a geographer with the U.S. Forest Service in Fort Collins, Colorado, and leader of the agency's Ecosystem Management Analysis Center. After joining the Forest Service in 1966, he earned a doctorate in physical geography from the University of California at Los Angeles in 1971. He is the author of *Ecosystem Geography* (Springer-Verlag, 1996). He has also published articles in journals devoted to geography, forestry, landscape and urban planning, environmental conservation, and environmental management.

Index

Page numbers with *f* indicate figures; page numbers with *t* indicate tables.